SpringerBriefs in Evolutionary Biology

For further volumes:
http://www.springer.com/series/10207

Clara B. Jones

Robustness, Plasticity, and Evolvability in Mammals

A Thermal Niche Approach

Clara B. Jones
Mammals and Phenogroups (MaP)
85 Merrimon Avenue, #12
Asheville, NC 28801
USA

ISSN 2192-8134   ISSN 2192-8142 (electronic)
ISBN 978-1-4614-3884-7   ISBN 978-1-4614-3885-4 (eBook)
DOI 10.1007/978-1-4614-3885-4
Springer New York Heidelberg Dordrecht London

Library of Congress Control Number: 2012937848

Printed on acid-free paper

Springer is part of Springer Science+Business Media (www.springer.com)

Preface

> *"[E]ach activity performed by an individual can be thought of as incurring a certain probability of death and a certain probability of successful reproduction ... ."*
>
> McCleery (1978)

> *"On a priori grounds the causation responsible for ... elimination [mortality] must be ecological or genetic."*
>
> Park and Lloyd (1955)

Robustness (relative insensitivity to system perturbation), plasticity (relative sensitivity to system perturbation), resilience ("stress–strain curve": X-axis = stress, Y-axis = response to stress = strain; pressure deviation= force/area), and elasticity (relative degree of resilience) have been fundamental concepts in applied physics at least since the eighteenth century (Malacarne 1783 in Rosenzweig 2007) and in the biological sciences at least since the early twentieth century when Cannon (1932; cf. JLF Bernard, *"thermodynamique"*) explicitly reconfigured the stabilizing bio-feedback construct (Schoener 2011; Bates and Cosintino 2011) as the mechanism, homeostasis ("homeorhetic" processes: Meiklejohn and Hartl 2002). Originally, use of the "homeostasis" concept among organismal and evolutionary biologists implied mechanistic "rheostat" processes used to characterize physical systems. Currently, emphasis is placed on the dynamic "balancing" of processes by negative and positive feedback mechanisms ("sensors") responsive to sensory communication (signals, information) about deviations from "target set 'points,'" including their thresholds, ranges, hierarchies, and probabilities of response. Many terms borrowed from physics have been "tweaked" by biologists, though mechanistic and reductionist approaches continue to define the scientific enterprise in the latter discipline.

Organismal processes, from molecular to higher levels, are effected by signal transmission (communication, "crosstalk": Laughlin and Sejnowski 2003) from one system unit to others (cellular differentiation, development: Heidel et al. 2011; Hallgrímsson and Hall 2005; cultural processes: Centola et al. 2011), events that partially explain differential life-history (energy allocation) tactics and strategies resulting from environmental information available to organisms (Kussel and Leibler 2005; Heidel et al. 2011). At every level of biological organization, feedback mechanisms are fundamental to resilience and elasticity since the former property describes post-perturbation return to energy balance while the latter property is the quantitative measure of resilience (Bates and Cosentino 2011). Under certain conditions, individuals may benefit from narrow "target set 'points,'" and under other regimes, broad "target set 'points,'" "mirroring" adaptive thermal tolerances maintained in dynamic equilibrium states by negative and positive feedback processes regulating an organism's optimal energy allocations to structures and functions. From the most basic levels of vertebrate organization, each biological unit can be dissected and diagnosed as a model of regulatory feedback responsive to within and between component responses to molecular or other stressors and strains (perturbations), dynamically optimizing and maintaining nonlethal levels of function.

I was introduced to these formulations in the 1960s during discussions with a neighbor who exposed me to the terminology of biomechanics, preparing me to integrate physical constructs and the study of animals (including hominids). Classes in economics and ecology interrelated the aforementioned formulations with additional ones, in particular, stability ("free [available] energy" states), complexity (multicellular, multinodal), buffering (regulatory mechanisms protecting phenotypes from environmental and genetic perturbation), canalization (low variability of a reaction norm), modularity (neural assemblies, synchronization of neurons), "norms of reaction" (phenotypic variation as a function of environmental variation), and flexibility (reversible responses). Lectures addressed niche theory (formulations about fluctuations in limiting resources, particularly, food), intra- and interindividual competition, thermal and biophysical effects, as well as allocation strategies and differential reproductive success of individuals, mean fitness of populations, and evolvability ("evolvability of a trait=coefficients of additive genetic variance, CV_a, + square of trait mean, l_a"; "evolutionary adaptability"). In the present treatment, evolvability will be defined as "the capacity to generate nonlethal heritable variation" (Kirschner and Gerhart 1998), an expression closely conforming to "thermal niche" effects.

Each functional unit of a biological system is characterized by an external surface or "phenotype" ("a periphery"). In the present review, "phenotype" refers to the whole organism's ("individual's") sensory-motor surface area exposed to abiotic and biotic stimuli external (exogenous) to the surface area. Where phenotypes of endogenous properties are intended (events internal to the whole organism), the particular "phenotype" of interest will be indicated. In nonpathological (nontoxic, nondisease, noninfirm, abnormal) states, the phenotype's reaction norms are expected to vary as a function of changes in thermal niches induced by environmental or other fluctuations, negatively impacting thermal tolerances and "free energy" states, often inducing compensatory responses (Ricklefs and Wikelski 2002; Vieira-Silva et al. 2011). Confronted with system perturbations deviating away from thresholds of nonlethal reaction norms, an individual's "physiology/life-history nexus" (Denver 1997; Ricklefs and Wikelski 2002; Price 2006; Vieira-Silva et al. 2011; Dantzer and Swanson 2011; Dell et al. 2011; Nevo 2001; Hau 2007; Stearns

1992; Hallgrímsson and Hall 2005) may be activated to enhance or diminish system sensitivity to a new regime, thereby connecting individual phenotypes and inclusive fitness with thermal properties of habitats (Dell et al. 2011; Nevo 2001). Modified fitness landscapes, then, may favor increased or decreased robustness and plasticity at one or more level of biological organization. The "regulatory logic" of reprogramming mechanisms is expected to change inclusive fitness optima under changed regimes, given endogenous (Kitazoe et al. 2011; Heidel et al. 2011) and exogenous (Slobodkin and Rapoport 1974) constraints, perturbations, stressors, and other interruptions of (thermal shock, "knockdown" events) and disturbances (risks, uncertainties) to system processes (Weinreich et al. 2006).

Aggregate effects of organisms' (synaptic) "decisions" partially determine shifting mean fitness of populations (Slobodkin and Rapoport 1974). In addition, individual tactics and strategies present as differential costs and benefits associated with modifications of reaction norms in response to altered patterns of environmental fluctuation. The persistence, reorganization, and collapse of biological systems depend upon differential energy inputs and outputs as well as scale, and lethal levels of imbalanced heat transfer may cause time-dependent, destabilizing cascades of system components (nodes, branches, circuits, networks, and modules) leading to system collapse (Bates and Cosentino 2011; Fig. 3.3).

Framing a selected review and discussion of robustness, plasticity, and evolvability in mammals was guided by the importance of these functional processes to differential reproduction of individuals and the production and maintenance of heritable variation in populations. The aforementioned topics are fundamental to the scientific study of basic and applied ecology and evolutionary biology (Schoener 2011; Ricklefs and Wikelski 2002) and are of current interest to many generalists and students, as well. Though this Springer Brief is not a specialized text and many of its examples and topics reflect my particular academic specializations, research, and interests, numerous references to the mainstream literature are provided to nonspecialists as maps to the rapidly expanding databases. In addition, an efficient treatment of structural, mechanistic, and functional aspects of the topics was required that did not limit the study's relevance to future speculations, insights, data, theories, and other quantitative treatments. Since I am not specialized in genetics and genomics, I was challenged to select citations and examples that were within my range of competence while, concurrently, retaining the specialist's engagement. I relied heavily on recent review papers, in part, for purposes of bibliographic economy, in part, to provide accessible, synthetic treatments to readers at any level of mastery. On the whole, citations were selected for their general applicability and accessibility to investigations at all levels of mammalian organization, and, wherever possible, I relied on Hallgrímsson and Hall (2005) that is treated, herein, as a single reference volume. The number of references is restricted, except in cases of serious controversy or interpretation. Bates and Cosentino (2011); Wagner (2001); Stearns (1992); Roff (2002); Chapters 6, 10, 11, 15 in Pigliucci and Muller (2010); Seebacher et al. (2010); Nijhout (2003a, b); Stearns et al. (1991); Rutherford (2000); and Hallgrímsson and Hall (2005) provide comprehensive introductions to many of the concepts reviewed in the present text.

Throughout, optimality terminology is employed as the *lingua franca* in physics, economics, ecology, and behavior (Slobodkin and Rapoport 1974; Giraldeau and Caraco 2000; Bates and Cosentino 2011; Wilkie and Godoy 2001). Even though "shifting optima" models are considered by some scientists to be more useful descriptions of populations in fluctuating regimes or in populations where "arms races" (coevolution: different mutations in more than one species acted on by selection, yielding similar phenotypes; Sect. 4.3, Box 3.1) strongly influence mean fitness of populations, these quantitative approaches are not as advanced as ESS treatments assuming frequency dependence. Furthermore, neuroscience research provides strong evidence for the importance of optimization processes in complex (multi-nodal) nervous systems (Vickery et al. 2011). Working definitions are provided parenthetically directly after a word or term's first usage unless wide agreement obtains in the nonspecialized biological literature for its use. Parenthetical citations are intentionally limited in number and format, with preliminaries (e.g., "also see," etc.) omitted for economy as well as visual appeal.

Common but slippery terms are differentiated, in particular, "adaptive" (a trait beneficial to a phenotype) from "adaptation" (an evolutionarily selected trait) and "variation" (realized differences within a population) from "variability" (whether or not character traits vary) after Gotthard and Nylin (1995) and Pigliucci (2008), respectively. Although the majority of extant mammals are eutherians, I have attempted to address monotremes and marsupials adequately. Where theoretical or empirical research on mammals was unavailable for critical issues (Frazier et al. 2006; Meyer et al. 2011; Tills et al. 2011), other taxa are referenced if the results involve extremely conservative traits (Meyer et al. 2011; Eschbach et al. 2011) or, less frequently, if the questions investigated appeared to apply to mammalian mechanisms without conserved features having been demonstrated (Coutinho-Abreu et al. 2011). Further, consistent with life table sums (Stearns 1992, Jones 1997a), female responses and parameters are emphasized throughout this review since male inclusive fitness is ultimately limited by the opposite sex (Emlen and Oring 1977; Trivers 1972, Wittenberger 1980) and by time (Schoener 1971), the only "truly fixed" parameter (Ricklefs and Wikelski 2002).

It is a pleasure to contribute to the SpringerBrief series, a format designed to fill an open niche for compact texts delivering efficient access to accurate information and productive ideas with the potential to promote learning and generate creative exploration. I am indebted to my Springer contacts, Melissa Higgs, Assistant Editor, Life Sciences/Biomedicine, responsible for copyediting my MS, and, especially, Janet Slobodien, Editor, Ecology and Evolutionary Biology, for her unparalleled, ongoing, patience and insight regarding all aspects of the project. I am honored that Stephen C. Stearns agreed to critically read my manuscript, and I benefited immensely from his comments. My assistants, Monica E. McGarrity (graphics), Jason Epperson (technical), and Anne R. Dachowski (bibliography), provided requisite skills and good humor. Tristan Burgess, Colin Chapman, Natalie Dawson, Carrie DeJaco, Ted Fleming, Brittany Grayson, Kayla Griffith, Frank Grutzner, James E. Mazur, Gary Mohr, Rick Murphy, Adam Reitzel, Elliot Tucker-Drob,

Polly Wiessner, and Gabriel Zunino generously responded to various inquiries related to mammalian ecology, evolution, diversity, and behavior.

Since graduate school, I have benefited from input and constructive criticism by professors, mentors, fellow students, colleagues, family, friends, and acquaintances. Among these, my dissertation advisor, the late William C. Dilger, Stephen T. Emlen, Jack Bradbury, Irwin S. Bernstein, Bernie Crespi, Iraneus Eibl-Eibesfeldt, the late John F. Eisenberg, the late R.F. "Griff" Ewer, Colin Groves, the late Harry Levin, Richard C. Lewontin, the late Jerry O. Wolff, Mary Jane West-Eberhard, OTS 73-2 instructors, the Japan Monkey Centre, and the German Primate Center deserve special acknowledgment. Clara K.J. Brown, a chemist, a high school science/math teacher, and my mother, was the first to tutor me in scientific research and design and, in no minor measure, was the initial stimulus for my sustained interest in botany, zoology, as well as research methods and design. A child's perceptions of the natural world are permanently modified by learning, for example, that trees can be identified by their leaves or by the configuration of their needles, that genotypes indirectly influence some phenotypic traits, or that the human species is energetically connected to other organisms, plant and animal, alike. Dalton A. Jones, Julie K. Palmer, and M. Luke Jones are exemplary adults for whom their mother is very grateful. The present review is dedicated to the late Jasper J. Loftus-Hills (1946–1974, cf. *Eleutherodactylus jasperi*), whose advice and support continue to inform and inspire my efforts.

Asheville, NC, USA   Clara B. Jones

Contents

Chapter 1
Introduction: What Paths to Inclusive Fitness of Individuals and Mean Fitness of Mammal Populations?

> *"The behavior of populations is an emergent property of the reactions of individuals to their circumstances ...."*
>
> MacDonald and Johnson (2001)
>
> *The reactions of individuals to their circumstances are emergent properties of the behavior of their populations.*
>
> cf. Schoener (2011)

Keywords Mammals • Life history • Robustness • Plasticity • Evolvability • Environmental grain

Stress is potentially lethal to animals, terminating future opportunities for direct or indirect reproduction. Within and between lineages, various mechanisms have evolved to mitigate deleterious effects of endogenous and exogenous perturbations to the whole organism (Nevo 2001). Within ecological windows (Schoener 2011; Hallgrímsson and Hall 2005), some stress responses, including novel ones, are heritable (mutation, epigenetic, hormonal), and others do not demonstrate effective levels (habits acquired by learning: Kawai 1965). A growing number of researchers (Badyaev 2005; Ghalambor et al. 2007; Espinosa-Soto et al. 2011) have concluded that, in principle, non-lethal effects, including novel ones, may facilitate adaptation if expressed recurrently on a mosaic phenotype comprised of traits evolving at different rates (Glanville et al. 2011; Huey and Kingsolver 2011; Porter and Kearney 2009; Dell et al. 2011; Suggitt et al. 2011; Hallgrímsson and Hall 2005, including, food restriction, food dispersion, food toxicity, social interaction rates, predator–prey, habitat architecture).

Using a regulatory circuit model, Espinosa-Soto et al. (2011) demonstrated that non-heritable, exogenous stimulation may rapidly induce heritable responses (mutation) and that phenotypic variations may be induced by phenotypes robust to mutation. Since robust phenotypes result in the buildup of mutations not expressed

C.B. Jones, *Robustness, Plasticity, and Evolvability in Mammals: A Thermal Niche Approach*, SpringerBriefs in Evolutionary Biology, DOI 10.1007/978-1-4614-3885-4_1,
© Clara B. Jones 2012

on phenotypes under "typical conditions," non-heritable, exogenous stressors acting on phenotypes robust to perturbation may induce novel, heritable variations ("genetic assimilation"). Espinosa-Soto et al. (2011) summarized their report thus: "Our observations suggest that phenotypic robustness to mutations can play a positive role in phenotypic variability after nongenetic perturbations," concluding that regulatory circuits are widely implicated in the expression of phenotypic novelties.

The model advanced by Espinosa-Soto et al. (2011) is formally persuasive, even elegant. However, other theoretical work suggests that evolution should favor "a reduction in environmental influences on phenotypic variation" and that "the phenotypes most strongly canalized against environmental perturbations are also those that are most strongly canalized with respect to genetic perturbations" (Meiklejohn and Hartl 2002). These results call into question the relationship between (non-lethal) environmental fluctuations (temperature, food availability) and robust phenotypes as Espinosa-Soto et al. (2011) describe it. For example, the 2011 model was "not sensitive to the magnitude of nongenetic perturbations." Modeling cellular differentiation, Siegal and Bergman (2002) showed a strong association between canalization and the complexity of Gene × Development interactions (cellular differentiation: Hirabayashi and Gotoh 2010). On the other hand, consistent with the circuit model of Meiklejohn and Hartl (2002), Sterck et al. (2011), studying components of forest dynamics, demonstrated empirically that responses to environmental stress depended upon relative degrees of change in intensity (I=power/area) from one stressful state to another rather than the absolute degree of intensity per se (efficient information transfer). Stearns and Koella (1986) highlighted another variable, duration of stressful events ("unavoidable stress"), suggesting that all states of "variation" (frequency, rate, duration, intensity, quality, type) should be considered when modeling Genotype × Environment × Phenotype effects, including near-lethal "knockdown" cascades (Christie et al. 2011) likely to occur in association with drastically stressful natural events, and the aforementioned studies suggest that stressors may manifest as the complete range of ways that features vary differentially.

Recent innovations in genome research demonstrating significant chromatin variation in mammalian genomes highlight the importance of clarifying the "plasticity" concept and of considering variation at multiple levels relative to Genotype × Environment × Phenotype effects in the Class (O'Brien et al. 1999). Comparative genomics has revealed recurrent, "core processes" at every level of biological organization (genetic circuits, protein regulation, neurotransmitter substances, biophysical patterns, Sect. 4.1), within and between taxa, characterized by both conserved and divergent elements (Denver 1997; Dantzer and Swanson 2011; Heidel et al. 2011; Cambridge et al. 2011). Similarities among organisms are thought to result from adaptive constraints at the molecular and, possibly, "intermolecular," levels as summarized by Weinreich et al. (2006, p. 113, Ricklefs and Wikelsky 2002): "It now appears that intramolecular [pleiotropic] interactions render many mutational trajectories selectively inaccessible, which implies that replaying the protein tape of life might be surprisingly repetitive." This statement and the comparative genomic database supporting it provide a quantitative foundation for general formulations of Genotype × Environment × Phenotype interactions, reinforcing the importance of qualitative and quantitative modeling efforts.

1.1 Mean Reaction Norms: "Mirroring" Spatiotemporal Variations in Microhabitat from Molecular to Phenotypic Levels of Organization

Genotype × Environment effects are manifested as an individual's mean reaction norms, representing responses to selection along a life-history "trajectory" ("maturation events": Stearns and Koella 1986). As one component of an animal's functional biology and as an ecological factor for other organisms, the phenotype is best considered not only in relation to gene (Kitazoe et al. 2011) and environmental (temperature, population density, habitat structure) factors, but also additional levels of organization (protein, physiology, cell differentiation) and properties of reaction norms linking all categories of animal organization with local ("patch," habitat) and global (population) regimes (Réale et al. 2003; Glanville et al. 2011; Rodriguez-Cabal and Branch 2011; Suggitt et al. 2011; Sterck et al. 2011; Sultan and Spencer 2002). Variations in energy allocation and resulting life-history strategies ("maturation events": rate of reproduction, age at maturity, longevity) may be positioned along a slow–fast continuum, from "slow development and long life span at one end and the opposite traits at the other end" (Ricklefs and Wikelsky 2002; McNab 1980).

For eutherians, energy expenditure is correlated with food habits (McNab 1986), whereby "fast" (relatively high metabolic rates, higher reproductive output) life histories are associated with a diet of, for example, "vertebrates, herbs, and nuts" and "slow" (relatively low metabolic rates, lower reproductive output) life histories, with "invertebrates, fruit, and the leaves of woody plants" (McNab 1986). The tripartite pattern of "maturation events" (energy expenditure, metabolic rate, reproductive output) is universal across animals, an observation leading Ricklefs and Wikelsky (2002; Weinreich et al. 2006) to infer that diversity of life histories are subject to similar constraints, probably physiological mechanisms (Gluckman et al. 2007; Glanville et al. 2011; Porter and Kearney 2009) responsible for cell differentiation (development) and other energy-allocation "decisions" determining inclusive fitness across individuals of a population, effecting shifting mean fitness optima. Studying phenotypic variability in larval amphibians, Denver (1997) proposed that vertebrate stress responses, in particular, those initiated by corticotropin-releasing hormone (CRH), are universal to the subphylum. Continuing the "search for shared mechanisms," Dantzer and Swanson (2011) investigated the relationship between the predictable features of vertebrate life histories and an insulin hormone (IGF-1), suggesting that endocrine processes may integrate traits associated with reaction norms. Both steroid (CRH) and polypeptide (IGF-1) hormones are critical for normal metabolic regulation, including thermal (cooling) and other physiological (glucose metabolism, stress-reduction) processes.

Popoli et al. (2011) reviewed glucocorticoid functions and stress responses in mammals. Glucocorticoids are a necessary component of feedback mechanisms associated with metabolism and stress-reduction and are implicated, as well, in cellular differentiation (development), cellular remodeling, and, most likely, regulation of gene expression in these circuits. Of further interest, glucocorticoids probably regulate the expression of insulin, a hormone complex functioning primarily to signal levels of glucose to an organism's cells. The role of glucocorticoids as a central promoter of stress–response circuits is further indicated by their capacities for rapid, non-genomic expression, nicely characterizing the sorts of ecologically "responsive" processes (including the functional, mosaic phenotype: Jones 2008) effecting reaction norms, including their variability and consequences, notably flexible age at reproduction (Suggitt et al. 2011; Adams et al. 2011; Christie et al. 2011; Slobodkin and Rapoport 1974; Stearns and Koella 1986). Using a geometric (biophysical) approach, Porter and Kearney (2009) studied the thermal niche of endotherms as a metabolic strategy (metabolic reaction norm) across climate gradients. According to the spatial (geometric) model devised by Porter and Kearney (2009), endothermy imposes significant energy and/or water costs that, similar to ectotherms, constrain an organism's survival, developmental, and reproductive "trajectories." These findings are consistent with theoretical treatments employed by Stearns and Koella (1986) whereby individual norms of reaction can be visualized as spatiotemporal "maps" of heritable phenotypic variation interacting with abiotic and biotic features of local regimes.

Four "take home" messages integrate and summarize not only this introductory section but also the whole review. First, all biological interactions, independent of analytical levels, are analyzable with a regulatory feedback model (upregulation–downregulation, excitatory–inhibitory "Hebbian" synapses, cultural mechanisms). Second, physiology and ecology are "mirrored" by quantitatively assessing thermal responses across organisms, including "unprecedented diversity" of features (traits, species, body sizes, habitats), revealing "generalities and deviations" that can be specified, quantified, and predicted (Dell et al. 2011), consistent with the recent report scaling metabolism universally for Class Mammalia (Hamilton et al. 2011). Third, across vertebrates, general patterns are derived from and constrained by similar biophysical responses to and interactions among endogenous and exogenous stimuli (thermoregulation, pupil dilation, "yawning," among others: Maynard Smith and Harper 2003), while "deviations" in mammals are primarily derived from and constrained by phylogenetic factors reflecting the causes and consequences of taxonomic diversification, including skeletal innovations subsequent to or preceding expansion of niches resulting from exposure to new food regimes (Glanville and Seebacher 2010). On the other hand, patterns of diversification may change if species ranges contract and/or if populations are extirpated, for example, by micro- or macroclimate change, habitat fragmentation, or shifts in limiting resources. Fourth, fluctuating endogenous and/or exogenous stressors may induce responses at one or more level of biological organization, from molecular to whole organism. Finally, an individual's relative (inclusive) reproductive (selfish) success is, ultimately, and, critically, a function of its thermoregulatory (economic, energetic, heat-transfer) responses to "risk of mortality" over time (Denver 1997; Ricklefs and Wikelsky 2002; Jones 1997a; Stearns 1992; Stearns and Koella 1986; Roff 2002; Slobodkin and Rapoport 1974; McCleery 1978; McNab 1980), counterpoised by negative and positive feedback functions (Cui et al. 2009). *In the final analysis, the present review intends to convince readers of the utility of viewing all biological events through the lenses of energy (E) and heat (Q), parameters, combined with time (T), capable of facilitating programs to quantitatively model events concerning shifting mean fitnesses of animal (including human) populations.*

Concepts, models, and datasets treated in the present review demonstrate multi-level interactions among system components, that each biological unit is characterized by a phenotype and an endogenous space, and that, even where initial stimuli originate in the environment external to the whole organism (temperature, limiting resource dispersion and quality), decisions to respond or not to respond are determined endogenously. Using a model from behavioral psychology: an initial stimulus (S) may induce a response (R: hypothalamus activation of the pituitary gland) that, in turn, becomes a second S inducing release of testosterone, a second R that may, in turn, ..., and so on. West-Eberhard's influential verbal models (West-Eberhard 2003, 2005; Jones 2006) acknowledge genetic "switch" mechanisms but omit explication of molecular, cellular, physiological, and other mechanisms that might have elucidated the nature of positive and negative feedback as regulators of the types of functional shifts detailed in the aforementioned example. It is possible that West-Eberhard's effective dismissal of the "homeostasis" concept in the early pages of her 2003 text led to oversight.

Chapter 2
Mammals: From Humble Vertebrate Beginnings to Global Terrestrial Dominance

"Patterson's material from Texas could simply be broken down in water and some of the processing was done in a creek on the site. Slaughter has set new standards in persistence, which have yet to be surpassed. At the Butler Farm locality he processed no less than 30 tonnes of matrix to obtain eleven teeth. Slaughter states that more than 200 tonnes have been processed in his search for Trinity mammals. This is a feat which speaks of itself; but only those who have actually carried out these washing and screening techniques truly appreciate its real magnitude."

Kermack and Kermack (1984)

"The radiation of the mammals provides a 165-million-year test case for evolutionary theories of how species occupy and then fill ecological niches."

Venditti et al. (2011)

Keywords Monotremes • Marsupials • Eutherians • Mammaries • Placenta • Energy investment • Epigenetics

Mammals are indicated for a review of robustness, plasticity, and evolvability because their morphology, phylogeny, biogeography, population and community ecology, and, particularly, energetics, ecology, life history, and behavior, are relatively well described. Further, laboratory studies, including experiments, of mice and other rodents have provided a large, detailed database on many aspects of mammalian biology including genes and genomes, biophysics, and pathologies. The relative degree of information available on mammals compared to any other vertebrate class permits a number of questions to be addressed that are likely to be widely applicable (protein diversification, life-history evolution, behavior). Notwithstanding many advantages associated with using mammals to investigate evolutionary patterns within and between vertebrate taxa, progress has been delayed

C.B. Jones, *Robustness, Plasticity, and Evolvability in Mammals: A Thermal Niche Approach*, SpringerBriefs in Evolutionary Biology, DOI 10.1007/978-1-4614-3885-4_2, © Clara B. Jones 2012

by the relative paucity of information about developmental and other trajectories coupling mammalian genomes and phenomes (Sect. 4.4).

A striking feature of the Class Mammalia is its worldwide ecological dominance and global success of this diversified group is a function of mosaic traits (characters evolving at different rates): endothermy (internal temperature regulation), mammaries (obligatory maternal care), placental structure and function (separation of fetal and maternal blood circulation), high cortex to brain ratios (sensation, perception, information-processing, learning, behavioral coordination and control), generalized, *totipotent* phenotypes (behavioral flexibility: Jones 2005a; Chap. 5), as well as features associated with local (patch, niche) specializations (frugivore–insectivore). A fundamental challenge to the life sciences is explaining the differential effects on phenotypes of endogenous (molecular, cellular), exogenous (climate), and epigenetic (DNA methylation) factors. Phenotypes may vary in time and space by chance alone and phenotypic drift may explain some mammalian features such as well-documented examples of phenotypic or "cultural drift" (Centola et al. 2011). The scientific study of mammals is currently "hot." Recent fossil finds, genetic and genomic research, and synthetic reviews have clarified and, in some cases, revised reconstructions of mammalian phylogeny, including, quantitative demonstration that metabolism scales equally across the class (Hamilton et al. 2011; Coda).

The mammalian fossil record is incomplete, in part, because many types arose in and dispersed through the tropics whose substrates are not propitious for the preservation of skeletal material. As a result, many questions remain unresolved and debated, and significant ambiguity exists about phylogenetic relationships within the Class, including the origins, differentiation, and trajectories of radiations, including the implications of Continental Drift (Eisenberg 1981, Figs. 2 and 4). Notwithstanding, based upon dental remains from ~100 Mya, it has been inferred that early mammals probably occupied insectivore–frugivore niches. Following Eisenberg (1981), Kermack and Kermack (1984), Vaughan et al. (2000), and Wilson and Reeder (2005), the Class Mammalia, air-breathing, warm-blooded (endothermic) vertebrates with females exhibiting mammary glands, is defined phylogenetically as all descendents of the most recent common ancestor of monotremes and placentals.

2.1 Robustness Matters: Appearance and Early Evolution of Mammals

After Kermack and Kermack (1984), synapsid mammalian ancestors diverged from the amniote line leading to reptiles at the end of the Carboniferous Period, ~300 Mya. Mammal-like reptiles ("non-mammalian reptiles") led to the first true mammals, precursors of extant members of the Class, between 250 Mya (Jurassic) and 100 Mya (Triassic) in the Palaeocene and Eocene epochs of the Palaeogene Period. Approximately 5,600 species of mammals are extant, classified in two subclasses, Prototheria (Infraclass Australosphenida, egg-laying mammals: Fig. 2.1) and the placental Theria (Infraclass Metatheria, marsupials and Infraclass Eutheria, non-marsupial placentals).

Fig. 2.1 Platypus (Monotremata, *Ornithorhynchus*). Conservation biologists have revived scientific interest in this ("solitary") species, recently recovered from near extinction. Differences in genetic and life history profiles; size, behavior, and pelage-color differences between populations; opportunistic foraging and dispersal habits; convergence with other venomous animals; and, seemingly flexible sociosexual organization exemplify noteworthy opportunistic plasticity in association with specialized and conserved biophysical traits (Grutzner personal communication). Sexual dimorphism is apparent in the species, although the role of sexual selection in platypus evolution has rarely been addressed, particularly, the possibility that the platypus "bill" serves as a display. *Ornithorynchus* and *Lemur catta* (Ring-tailed lemur) exhibit some convergent characteristics (limb "spurs": Groves personal communication), however, this area of investigation remains unexplored

Mammalogists studying fossil and extant types utilize a variety of traits for diagnostic purposes. Kermack and Kermack (1984), paleontologists, proposed a list independent of soft tissues. These authors identified the following as universal traits (1) 2.3.3.3.3 phalangeal formula, (2) hair or fur (pelage), (3) milk glands in females used to nourish young and obligating that sex to parental care, (4) details of jaw articulation, (5) three bones connecting the tympanic membrane to the inner ear, (6) some teeth with more than one root, and (7) a four-chambered heart. In their introductory textbook, Vaughan et al. (2000, Table 2.2) provide an expanded list of diagnostic characters. Kermack and Kermack (1984) concluded, however, that only two characters unambiguously differentiate mammals from other taxa (1) the presence of three inner-ear ossicles and (2) a "squamosal-dentary" jaw-joint. Following Eisenberg (1981), both traits, in addition to modifications in phalanges and dentition, concern feeding tactics and strategies, in particular, refinements of sensory

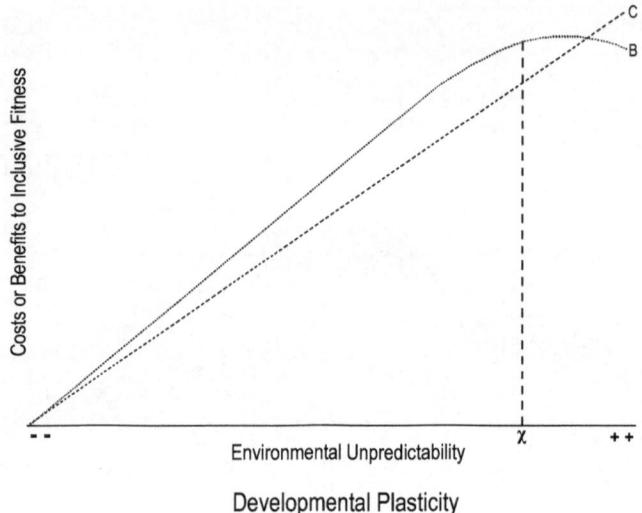

Developmental Plasticity

Fig. 2.2 Environmental stressors sufficient to destabilize regulatory feedback processes inherent to all biological processes may arise from numerous sources (colonization, toxicity, drought, increased or decreased temperature), challenging organisms to express compensatory responses. Presented is a simple graphical model of the relative costs (C) and benefits (B) to inclusive fitness of developmental plasticity as environmental unpredictability increases from low (*double minus*) to high (*double plus*). Benefits increase to asymptote, leveling off as costs increase linearly (for limiting resources), and the maximum net benefit ($B - C$) should occur at x, where benefits maximally outweigh costs, a function of environmental unpredictability over the short and long terms. In nature, x will manifest as the mean of a range of response norms, including response hierarchies and transition probabilities, rather than a single point. Note the expected intersection of cost and benefit curves such that, wherever plasticity is initially beneficial or becomes beneficial, costs to lifetime inclusive fitness, eventually, outweigh benefits where benefits reach asymptote (limiting resources are depleted below levels required for reproduction). See Discussion for further analysis of this graph. ©Clara B. Jones

capacities and capture techniques during co-evolution of predator–prey relations. The fossil record also documents wide variations in body size among mammals, additional features suggesting dynamic intra- and intertaxonomic interactions directly related to their highly differentiated roles in ecosystems, intra- and interspecific competition for limiting resources, and the global dominance of non-volant, terrestrial mammals since the Pleistocene.

Many extant branches of mammals have diversified in thermal niches that do not "fill up" ("open" niches) because of temporally heterogeneous conditions in which ecological factors operate as a "moving target" (Venditti et al. 2011; Fig. 2.2). Depending upon rates of environmental fluctuations relative to generation times, "open niche" regimes are likely to favor mechanisms permitting "responsive switching," a feature of neural plasticity characterized by facultative physiological and behavioral changes in response to fine-grained environmental fluctuations (Kussell and Leibler 2005). "Moving target" environments (Fig. 2.3; Box 2.1; Chap. 3) are

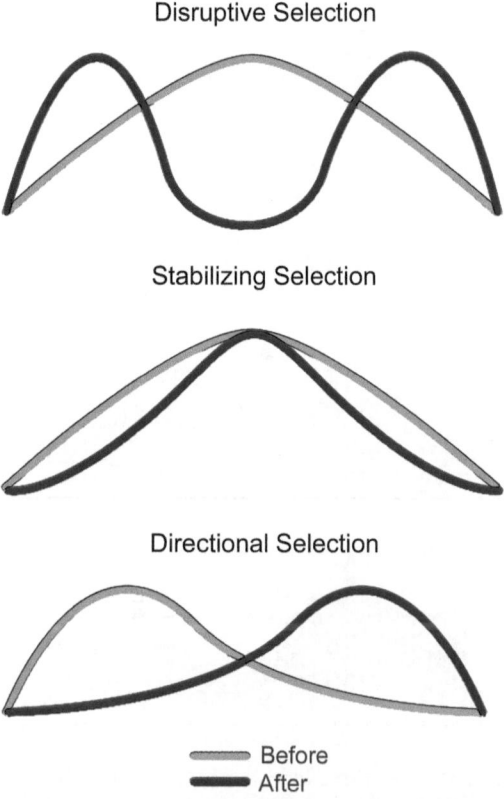

Fig. 2.3 As Rutherford (2000) put it, "Evolution is generally modeled as the conversion of genetic variation between individuals into genetic variation between populations." "Conversion" processes may have more than one consequence, including (*top*) Disruptive or, Divergent Selection whereby heritable and/or nonheritable phenotypic variation within a population is subdivided by reproductive isolation (Chap. 4), (*middle*) Stabilizing Selection ("selection for intermediacy": cheek teeth of herbivorous mammals) whereby genotypes around the central tendency of a distribution of genotypes in a population are more reproductively successful than genotypes at the tails of a distribution (many conservative or polymorphic traits), and (*bottom*) Directional Selection whereby genotypes at one or another tail of a distribution of genotypes in a population are more successful reproductively ("rapid evolution": Sect. 4.2). Many cases of "rapid evolution" are explained by (strong) Directional Selection that might, for example, be induced by perturbations (stressors) with the potential to insult a whole organism (Chap. 4). By definition, a genotype's lifetime (and beyond) reproductive success obtains relative to other genotypes in its population

strongly correlated with mammalian diversity, presenting some groups with a broad range of adaptive opportunities (Venditti et al. 2011; Gibb et al. 2011).

Analyzing molecular phylogenies for mammalian families, Meredith et al. (2011) showed that interordinal diversification was initiated by ecological events occurring during the Cretaceous Terrestrial Revolution ~125–80 Mya, in particular, the rapid rise of angiosperms (prey). Intraordinal expansion, on the other hand, was coincident

Box 2.1

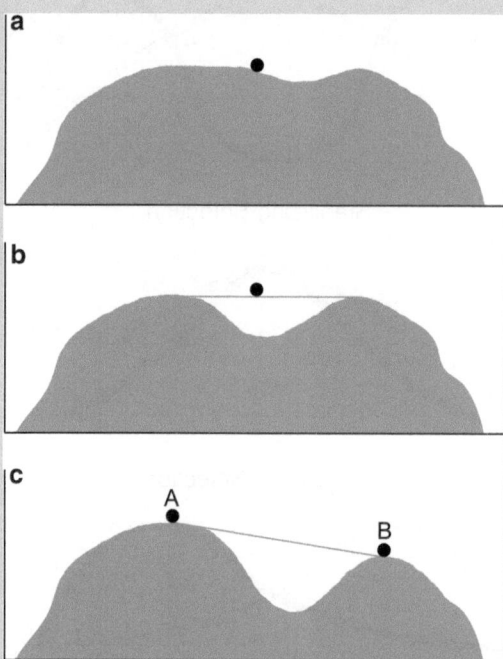

This illustration represents three landscapes (a, b, c) with different conformations: (a), a relatively undifferentiated spatial landscape, (b), a slightly differentiated spatial landscape, (c), a highly differentiated (divergent) spatial landscape. Each black ball represents optimal reproductive success for one genotype (reaction norm). In conditions "a" and "b," a single genotype achieves optimal fitness. In other words, the same genotype is favored, on average, in each "state" within the landscape (or "patch"). Condition c displays two optima favoring two different genotypes. In condition c, two environmental "states" have arisen, possibly due to a mutation more efficient in one state than another or because individuals of one genotype (a group of kin or a population subunit dominated by one prolific male) are more successful exploiters of resources clustered in one state compared to the alternative genotype. Landscapes b and c schematize heterogeneous "states" whereby mammals are likely to experience >1 environmental "state" across its lifetime (multiple prey types, multiple predators or parasites, multiple mates). Note that in Box 1b, the genotype has slightly higher reproductive success in one area of the landscape, and in Box 1c, genotype A experiences relatively higher reproductive success than genotype B. Following Jones (1997a; Gaillard et al. 2003), from the perspective of most mammals, "states" in which populations are located may be described with an "autocorrelation function" expressing the relationship between

(continued)

Box 2.1 (continued)

environmental events (temperature, rainfall, phenology) and some unit of time (month, year). No or negative autocorrelations between event and time period provide organisms with no reliable signals (information) upon which to base life history decisions (gestation, lactation, dispersal), while positive autocorrelations indicate environmental predictability. Related to decision-making in heterogeneous regimes, the important phenomenon of "lag" between environmental events used by mammals to inform responses mediating life history "trajectories" deserves prioritization as a subject for research.

In a "coarse-grained" environment, individuals of a population experience the same "state" across its lifetime, *ceteris paribus*, although, by definition, mean reaction norms are direct functions of individual genotypes. The researcher expects empirical results to vary by sex under the same regimes for a given population of mammals, mean BMR of males will be higher than that of females. By inference, these sexually dimorphic states predict that, on average, male physiological processes will be more efficient than those of females, so that males will have more "free energy" than females to invest, if required, in flexible responses to extreme stress. Perturbations on female physiology, on the other hand, are expected to induce more extreme stress responses compared to males in the same regimes, yet females, because of relatively lower BMR, will have less "free energy" to respond flexibly to extreme stress. This female paradox (greater susceptibility to stress but, at the same time, less "free energy" to respond to the challenge) has, apparently, favored more canalized female phenotypes over evolutionary time, consistent with empirical results (Lerner 1954) and verbal modeling (Jones 2005a, Chap. 6). Males may not "do it better," but males do it faster, with consequently higher variation in reproductive success and greater potential lifetime reproductive success compared with females in the same population. Several conclusions obtain from sexual dimorphism in thermoregulation and thermal tolerances in mammals, such as higher costs or fewer benefits to males from compensatory indirect reproduction, predicted and observed empirically (females more "social" than males: Jones 2005a, b, Fig. 1). Other costs and benefits obtain to males such as greater "genetic load," discussed recently by Agrawal (2010).

An environment is said to be "fine grained" when fluctuations impact individuals within their population's generation time (mean lifespan or time required for population to double in size; Gaillard et al. 2003). For example, in one Costa Rican mantled howler monkey (*Alouatta palliata*) population, generation time equaled 6.27 years and rainfall from the site exhibited positive autocorrelations at 3 and 6 months as well as 1 year. Each of the predictable cycles is briefer than generation time for this population ("fine-grained" conditions), the 6-month autocorrelation was strongest, and significant features of life history reflect the most informative signal (6-month gestation length). Even the most specialized mammalian taxa exhibit noteworthy behavioral flexibility

(continued)

Box 2.1 (continued)

(red panda: *Ailurus fulgens*, giant panda: *Ailuropoda melanoleuca*, nine-banded armadillo: *Dasypus novemcinctus*), a characteristic of highly conserved mammals, as well (platypus: Sects. 2.1 and 2.2). For mammals and other groups (plants, insects), confident prediction of specialist or generalist life history strategies has been elusive, possibly because mammals display minimal propensity for habitat selection, suggesting that models should be based on microclimate data (Chap. 3). In addition, organisms confront differential costs and benefits whether they experience "coarse-grained" or "fine-grained" "states," and the observation that most mammals encounter some combination of both conditions across their lifetimes increases analytical complexity. Accessible verbal and quantitative treatments of the foregoing topics are available in Emlen (1973) and more recent treatments (Baythavong 2011; Chap. 3). ©Clara B. Jones

with increased availability of "ecospace" associated with "distinctive morphological adaptations" resulting from the Cretaceous–Paleogene mass extinction. The same authors point out that rates of molecular evolution varied (fast rodents, slow cetaceans), constrained both temporally and topologically, patterns of diversification probably reflecting differential grand mean fitness optima of populations within species. Investigating animal diversification patterns with fossil and molecular evidence, as well as developmental data and descriptions of feeding strategies × habitat, Erwin et al. (2011; Fig. 2.2) identified a macroevolutionary "toolkit" among animal clades originating ~850–635 Mya and pulses of diversification among metazoans ~635–488 Mya. The success of metazoans was associated with variations in ecological webs and biodiversity preceded by novel patterns of developmental regulation.

Soria-Carrasco and Castresana (2011) documented the important role of ecological factors for reconstructions of mammalian diversification, and Venditti et al. (2011) demonstrated strong correlations between fluctuating niches and morphological diversity, apparently coupling heterogeneous environments and origins of biophysical novelties. Studying a highly diverse bat Family (Phyllostomidae), Dumont et al. (2011) reported that dietary evolution (changing patterns of food exploitation) was highly correlated with biophysical features, in particular, variations in skull morphologies and their relationship to biting indicators within and between species. Consistent with the aforementioned reports, speciation rates in these bats were preceded by stabilization of morphological characters. Additionally, the most conserved and derived species exhibited the most extreme morphologies and the highest levels of frugivory, implicating, again, a fundamental morphological toolkit "linking phenotype to new niches through performance," energizing niche-based speciation rates. Similar results for other vertebrate taxa reinforce inferences that a general "regulatory logic" underlies variations in niche characteristics and variations in species abundances and distributions (Greenstreet et al. 1998; Ingram and Mahler 2011).

In biological systems, specifications of conserved traits, including their origin(s), structure(s), function(s), divergence ("transitions"), maintenance, diversification, and variability, usually serve as assays for robust features, primitive "toolkits," circuits, and networks regulating functional biological diversity [genetic adaptation for "rapid evolution": Christie et al. (2011), plastic rules for chromatin organization and protein switch mechanisms: Tsankov et al. (2011), functional protein turnover across mammals: Cambridge et al. (2011), within-taxon genome differentiation in mammals: Keane et al. (2011), human protein evolution: Cui et al. (2009), evolution of gene expression in mammalian organs: Brawand et al. (2011)]. Though the particulars of trajectories from molecular to phenotypic levels vary within and between taxa and are not well dissected, Christie et al. (2011) concluded that regulatory feedback mechanisms driven by intra- and intercellular communication demonstrate "a remarkable versatility" in creating and exchanging small molecules. General programs stabilizing plastic expressions of biological units are discussed in greater detail below (Chap. 3, Sect. 4.1).

2.2 Order Monotremata: Good Work Ethic, Abundant Food, Lots of Luck

Most of the details in this and the following section are extracted from Grutzner (personal communication), Eisenberg (1981), Grant and Temple-Smith (1998), Vaughan (1972), Vaughan et al. (2000), Feldhamer et al. (2004). All of these authors emphasized the incomplete nature of the mammalian fossil record, with a high degree of uncertainty surrounding generalizations. A *Journal of Mammalogy* search yielded only 12 citations for the egg-laying monotremes, not all of these technical reports. Numerous articles on monotremes (Tachyglossidae: echidnas [spiny anteaters], Australia, Tasmania, New Guinea; insectivores) and (Ornithorhynchidae: duck-billed platypuses, E. Australia, Tasmania; carnivores) are found in specialized biology journals, primarily because of (1) conservative (reptile; avian: egg-laying; olfaction), derived (electroreception, phalanges), and ambiguous (skull morphology) traits, making these families a model for the historical reconstruction of events leading from reptiles to birds to mammals, (2) the broad import of platypus for studies of epigenetics (Fig. 2.4), genomic imprinting, and sexual determination, (3) echidna's importance as subjects in research on sleep, (4) the importance of monotremes for conservation biology (population recovery after translocation), (5) behavioral and genetic differentiation between populations (pelage color, size, behavior), and (6) their interest for investigations of behavioral thermoregulation. Relative to body size, all monotremes are long-lived with a low reproductive rate, producing one or two altricial young yearly, features associated with relatively long life spans and extended, often elaborate, parental care ("slow" life-history trajectories).

Fig. 2.4 Epigenetics involves changes in gene expression levels induced by effects on "small [regulatory] RNAs" and, in mammals, transposable elements (TE) may mediate epigenetic regulation (histone modifications) (Huda and Jordan 2009). This figure displays laboratory mice (*Mus*) from the same clone exhibiting different phenotypes induced by different environmental effects (diet) on developmental "trajectories," effects that are not necessarily straightforward (Gluckman et al. 2007). The relative significance, singly and in combination, as well as their consequences for individuals, populations, and habitats, is in the early stages of investigation for animals. Epigenetic effects may be (developmentally) time dependent ("critical" or "sensitive" periods), are fundamental to studies of sexual dimorphism in the mammalian nervous system, including "genomic imprinting," and potentially influence functions at all levels of organization, both nonpathological and pathological (Nugent and McCarthy 2011; Francis et al. 2003; Day and Sweatt 2011). Detailed treatments of epigenetics are beyond the scope of this review; however, two current areas of research with fundamental relevance to mammalian genomics concern the evolutionary import of epigenetics and its heritabilities across taxa and conditions

Early investigations of monotremes have been validated by recent studies (Grant and Temple-Smith 1998; Eisenberg 1981; Grutzner personal communication) expanding our knowledge of this Order. Except for the obligatory mother–young association, sexual segregation and female intolerance of males obtains outside the mating season, comparable to most mammals. Grant and Temple-Smith (1998) confirmed that, despite adaptations to strongly seasonal regimes, variations in burrow complexity within and between families, and the (apparent) display of incipient sociality (numerous anecdotal and a few empirical reports of variegated maternal behavior in both families, occasional polygyny, aggregations, allogrooming between platypus mates, male dominance hierarchies during breeding season among echidnas), no species in the order has become truly fossorial, a protective–defensive trait often associated with higher grades of mammalian sociality.

The carnivorous platypus (Fig. 2.1) exemplifies interacting effects of morphology, dietary niche, and allocation strategies (Grant and Temple-Smith 1998; Eisenberg 1981; Ewer 1968). This species feeds primarily on benthic invertebrates (shrimp, insect larvae) found in streams subject to seasonal desiccation, a recurrent ecological perturbation inducing hormonal and related stress responses, including increased exploratory responses such as long-distance movements (Fig. 2.2). Platypus may increase niche breadth by switching opportunistically between food items depending on fluctuating conditions such as water temperature, variable rates of water flow, streambed quality, and food availability or abundance, factors, among others, explaining their restricted geographical distributions, historically and presently. *Ornithorhynchus* phenotypes are characterized by independently evolving (mosaic) traits permitting noteworthy capacities to conditionally vary niche breadth, though the species is constrained evolutionarily by extremely derived traits such as a large, exaggerated "bill" that may have been favored by sexual selection as a display of male quality. This idea is supported by observations that the species appears to switch conditionally from a solitary to a polygynous mating system (Grant and Temple-Smith 1998).

On the other hand, minimal sexual dimorphism is evident in echidna and platypus, though males of both families exhibit a "horny spur" on the hind foot, specialized in platypus for chemical communication via functional venomous glands. Eisenberg (1981) reported that the functional anatomy of male platypus spurs is "unique to living mammals," but a spur also characterizes male ring-tailed lemurs (*Lemur catta*: Groves personal communication), apparent convergences worthy of comparative study. The platypus bill, and additional specialized structures and habits (webbed, clawed phalanges specialized for burrowing, nest-building, and food capture), may have been favored by directional selection and "rapid evolution" (Fig. 2.3), resulting from population bottlenecks (Grutzner personal communication) and subsequently limited access to resources. Importantly, the platypus bill is specialized for substrate feeding, adaptations consistent with life along streambeds and a diet of benthic organisms (McNab 1980). Presently, and historically, *Ornithorhynchus* exhibits a limited geographic range and restricted diversification, patterns of distribution unlike pelagic-feeding fish (Greenstreet et al. 1998; Ingram and Mahler 2011) and mammals (Pyenson and Lindberg 2011), species groups occupying a broad range of ecological niches. Numerous questions related to proximate and ultimate causes and consequences of monotreme biology await investigation.

2.3   Subclass Theria (Infraclasses Metatheria and Eutheria): "Tinkering" with Contents of a Generalized "toolkit"

Therians are widely distributed (Metatherians: Australia and surrounding areas, Neotropics; Eutherians: global terrestrial, volant, water). Most metatherians (Fig. 2.5) exhibit a "fast" life-history strategy (short gestation periods, semelparity), although young are relatively precocial, using strong arms for transport into the mother's pouch ("marsupium"). According to Eisenberg (1981), the foregoing features may represent adaptations to xeric environments. Similar to monotremes, most eutherians exhibit long gestation periods, iteroparity, and altricial (relatively helpless) young requiring extended periods of maternal care.

Fig. 2.5 Koala (Diprotodonta, *Phascolarctos cinereus*), the only extant member of the Family Phascolarctidae, is found in Eastern Australia, feeds on the leaves of a few eucalyptus species, and, exhibits a "slow" life history strategy, contrary to the pattern of most marsupials. Features of energy investment in koalas are related to the very high costs associated with specialized folivority in addition to the demands of survival in extreme environmental regimes. Though characterized by sexual segregation, male–male aggression and temporary male dominance hierarchies have been observed. Marsupials have figured prominently in recent scientific reports about mammals (genome sequencing, evolution of placentation, evolution of social behavior and auditory communication, conservation biology). *Phascolarctos* is a "flagship" (popular, high-profile) species whose conservation may maximize likelihoods of preserving associated biodiversity

McNab (1986, 2006), in particular, studied the implications of eutherian energetics relative to inter- and intraspecific competition. The previous author found that "conservative" mammals specialize on foods requiring low rates of basal metabolism (BMR: f body size × temperature × diet), escaping competition with "advanced," low-BMR mammals with a wider thermal niche (Chap. 3; Box 2.1). McNab (1986, 2006), also, observed that low-BMR mammals are particularly vulnerable to the effects of competition and predation because of the high, usually, obviating, costs required to increase reproductive rates ($r \rightarrow$ population fluctuations), and, high-BMR generally outcompete low-BMR mammals, an effect that is a function of food type. Related to the former finding, McNab (2005) highlighted the observation that, in general, herbivorous eutherians require water. The previous author argued that long-term survival of low-BMR mammals depends upon their "isolation" from predation and competition (islands, release from eutherian predation [carnivores], areas of low species diversity, "isolation from the general fauna").

It is possible to assess McNab's (1986, 2006) propositions relative to the evidence about patterns of mammalian evolution detailed in the previous citations in this section as well as the summary of differential competitive relations among Theria in Vaughan et al. (2000) (1) Eutherians have radiated more widely than metatherians, (2) The generalized, metatherian "structural plan" is more conservative than that of eutherians. Importantly, Eisenberg (1981; McNab 2005) took the position that this feature promotes mammalogy by serving as a "control group" for other taxa. (3) The range of body masses is less in metatherians. (4) Metatherian social systems are not particularly differentiated compared to eutherians, although, it might be added, there is a not insignificant diversity of social organization among marsupials.

(5) A fifth difference between metatherians and eutherians noted by Vaughan et al. (2000; Elliot and Crespi 2006) is the greater taxonomic diversity of the latter compared to the former Infraclass. (6) The eutherian cortex develops faster than and is relatively larger than that of marsupials. Importantly, this metatherian characteristic obtains because of their reproductive physiology (Padykula and Taylor 1982), particularly, an inefficient placenta compared to the "true placenta" of eutherians. The eutherian foetus experiences a longer placental phase compared to metatherians, dedicated to a "fast" allocation strategy due to the truncated period of intrauterine development. It is of interest that monotremes, particularly, *Ornithorynchus*, feature relatively large and "complex" brains, reflecting their reproductive strategy and the egg's protective environment, permitting a relatively long period of pre-hatching development. The monotreme and eutherian cases highlight the pattern of greater investment by females in gestation over lactation, the most common pattern among mammals. Interestingly, however, marsupial ("fast" life history) and primate ("slow" life history) females invest more in lactation than gestation, phenomena discussed below (Chap. 3). (7) Eutherians exhibit greater phenotypic variability than metatherians, a finding consistent with the latter taxon's canalized "structural plan." (8) Endothermy is more efficient in the offspring of eutherians. Vaughan et al. (2000) advanced this feature as a beneficial factor; however, it is worth pointing out that endothermy comes at a high energetic cost that must be borne by the eutherian female, except for the very uncommon cases of *significantly beneficial* paternal investment in the Class (Callitrichidae). In the mammal literature, paternal associations with immatures are generally reported as behavior beneficial to young, and costs are rarely weighed. All of the foregoing contrasts may be posed as hypotheses, submitted to quantitative modeling, descriptive and literature investigation, and experimentation.

2.4 The Mammalian Placenta: The "Intimate Connection" Between Female Reproductive Physiology, Offspring Heat Regulation, and Life-History Strategies

In addition to mammary glands, one of the most conservative mammalian features, the eutherian uterus, represents a major evolutionary transition. After Eisenberg (1981), eutherians are characterized by a "true" (chorioallantoic) placenta permitting an "intimate connection" between foetus and female via specialized villi, barriers to the transfer of substances. This system affords a trade-off between high energy costs to the mother from an extended period of gestation and the benefits of long gestation (rapid cortical development), an allocation trade-off associated with wide geographic distributions in some taxa, topics needing intense investigation. Continuing to follow Eisenberg (1981), the metatherian (choriovitelline) placenta lacks villi, preventing an "intimate connection" between female and young, a structural pattern that, inevitably, leads to increasingly negative effects from leucocytes, eventual induction of an "immune-rejection response" in females, and parturition, events constraining energy-investment in offspring. Bandicoots (Peramelemorphia), terrestrial, omnivorous, marsupials, exhibit a eutherian-like placenta. Research on negative and positive feedback regulation of temperature ↔ female physiology (mammaries, ovaries, placentas, mitochondrial function) ↔ ecological (diet) requires prioritization (Dell et al. 2011; Seebacher et al. 2010).

In impressive detail, Blackburn and Flemming (2011) reviewed literature documenting primitive placentation patterns characterizing all live-bearing lizards, describing a newly identified placental skink, *Trachylepis ivensii*. Lizard placentas exhibit an embryo retained in a remnant egg shell, nourished by an egg yolk, an allocation program limiting gestation time and enhancing post-parturition energetic costs to females. A few live-bearing lizard species gestate embryos in small eggs with minimal contribution of nutrients from the placenta. The "true" placenta identified in *T. ivensii* allows implantation of embryos in the oviduct wall, retaining a yolk but nourished primarily by the mother via the placental wall whereby the embryo is in close contact with the female's blood vessels, displaying the most intimate association of this sort among lizards. However, contrary to the eutherian placenta, the association between *T. ivensii* fetus and female does not avoid immunological problems, including the possibility of male embryo feminization. Two additional skink species display similar mechanisms of embryonic nourishment, and Blackburn and Flemming (2011) call for these taxa to be classified as "true" placentals.

Chapter 3
Variability of Mammalian Thermal Niches: Differential Effects of Local and Global environmental Heterogeneity

> "Both transcriptional mechanisms and membrane composition interact with environmental temperature and diet, and this interaction between diet and temperature in determining mitochondrial function links the two major environmental variables that are affected by changing [micro- and macro-] climates."
>
> Seebacher et al. (2010)
>
> "Each group of vertebrates has its own characteristic history of evolutionary adaptation to new habitats and life strategies. If this history is preserved in mitochondrial (mt) genomes, a comparative study across various vertebrate groups may detect new aspects of mt functions and answer the long-standing question of whether large variations in mtDNA sequences are due to adaptive evolution of amino acid sequences or nucleotide mutation pressure."
>
> Kitazoe et al. (2011)
>
> "It is usual to speak of an animal as living in a certain physical and chemical environment, but it should always be remembered that strictly speaking we cannot say exactly where the animal ends and the environment begins—unless it is dead, in which case it has ceased to be a proper animal at all ...."
>
> Elton (1936)

Keywords Energy • Thermal niche • Thermal tolerance • Metabolism • Microclimate • Seasonal tropical forests

Since the discipline's inception, ecologists have studied spatiotemporal variations in thermal tolerances of individuals and their reaction norms within and between populations (Shelford 1911). Events with the potential to induce heritable (genetic variance: phenotypic variance ratio: Meiklejohn and Hartl 2002) variation in populations have been summarized by several authors: Hallgrímsson and Hall 2005. Additional sources of within-population variation are easily advanced: transcription errors, ritualization, perturbations of a wide variety, dynamic (mitochondrial) effects

C.B. Jones, *Robustness, Plasticity, and Evolvability in Mammals: A Thermal Niche Approach*, SpringerBriefs in Evolutionary Biology, DOI 10.1007/978-1-4614-3885-4_3, © Clara B. Jones 2012

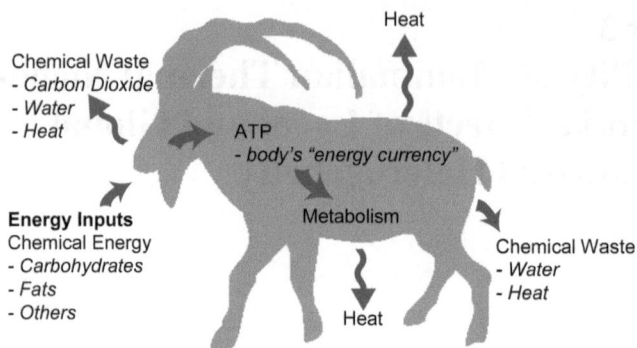

Fig. 3.1 Endotherms (birds, mammals) generate heat (Q) endogenously for maintenance and regulation, within a nonlethal range, of body relative to ambient temperature ("thermoneutral zone": TNZ, Sect. 3.1). Metabolism entails those intracellular processes required for the conversion of food into energy (E: a form of heat) used to sustain life. Adenosine triphosphate (ATP), an intracellular compound produced by mitochondria (respiratory and E-producing organelles with their own nucleus), transfers E for metabolic functions (Seebacher et al. 2010). ATP production and metabolism are regulated by feedback processes sensitive to the ratio of ATP and other cell products. Asymmetries, and, other, recurrent discontinuities, provide information to and effect communication within and between all levels of organization within an organism, and are, at least, in theory, measurable variables from which quantitative relationships (ratios) can be determined, providing information used by cells to estimate and predict organism functions relative to external conditions. It is a relatively trivial exercise to express ratios reflecting differences that, when measured over time ($T_0…T_1…T_2…$), provide information about temporal changes (ΔT) in differences between variables of interest (ratios: ATP:AMP, temperature:nutrient availability). All regulatory biological processes function in accord with this fundamental calculus (comparison, contrast, matching: Gutman 1977), and Seebacher et al. (2010) pointed out that variations in diet and temperature (proximate factors) are primary determinants of mitochondrial function (proximate factor), in addition to, it might be added, mean reaction norms (proximate factors), survival (ultimate factor), and reproductive success (ultimate factor). See de Jong (1976) and Chap. 3 for further details and discussion. ©Clara B. Jones

(Fig. 3.1), and, combined with research on regulatory mechanisms, are active fields of investigation (Huda and Jordan 2009; Chap. 3; Sect. 4.1, Discussion) in genetics and genomics. Porter and Kearny (2009) used biophysical models to quantify the selective value of functional traits (phenotypic variation) in mammals. According to these authors, spatiotemporal variations in traits associated with energy inputs and outputs correlate with variations in local ("patch," microclimate) and global (population, macroclimate) conditions.

The method employed by Porter and Kearney (2009) clarified mechanistic parameters of energy and water exchange ("distributed heat generation") determining developmental constraints realized as minimal and maximal ranges and thresholds of phenotypic characters as well as norms of reaction. Porter and Kearny (2009) described the "comfort zone" of mammals and other endotherms as the local ranges and thresholds of "critical" stimuli (temperature, wind speed) beyond which metabolism (energy and water balance) is perturbed, in particular, mechanisms of heat production and loss. The "biophysical" outcomes of fluctuations in these processes will be related directly to an organism's "mean norms of reaction," survival, and reproduction, as well as differential selective values of traits, including their variation. The organism's "comfort zone," then, is that multidimensional and changing space in which energy and water balance sustain reproductive and life functions. This thermonuclear zone (TNZ), including its stochastic features, determines not only the biophysical traits encompassed by Porter and Kearney's (2009) model (plasticity, pelage color, body size, morphology; Sect. 4.1) but also what Stearns (1992) discussed as the "extrinsic age- and size-specific shifts in mortality rates that interact with...the intrinsic constraints and potentials of organisms." The present review concerns the aforementioned topics with specific emphasis on robustness, plasticity, and evolvability related to all levels of abiotic and biotic organization endogenous and exogenous to the organism.

3.1 Are There General Laws of Mammalian Thermal Niches and of Thermal Tolerance Evolvability?

["Homology of germinal cause"] is now apparent between *Mus musculus* and *Rattus norvegicus*, which have varied so far from a common type that they are now inter-sterile and have been placed recently in different genera. Yet they have retained a genetic construction so similar that it contains genes common to both species. Whether this is due to a community of descent in the terms of current evolutionary theory or to relationship through some other cause is one of the questions which genetics, aided by the chromosome notation may be expected at some time to answer. L.C. Dunn (1921)

Energy metabolism is temperature sensitive, and animals respond to environmental variability at different temporal levels, from within-individual to evolutionary timescales. Seebacher et al. (2010)

Cooper et al. (2011) investigated phylogenetic niche conservatism (PNC) in mammals. PNC, a state in which some closely related taxa occupy similar niches, is related to species distribution patterns influenced by ecological factors, and, it might be suggested, may result from clustering of such taxa for efficient access to one or more clumped, unpredictable, rare, and/or ephemeral resource. Cooper et al. (2011) hypothesized that species predisposed to high PNC would evolve at slower rates than species exhibiting low PNC due to their retention of conservative, less "labile," and less variable thermal niches. These same authors predicted that taxa characterized by less conserved thermal niches would accommodate more successfully to environmental stressors since ranges and thresholds of "environmental tolerances" are heritable. These variables relate directly to costs and benefits of plastic responses for individual survival and reproductive success as well as shifting population optima.

PNC resulting from ecological similarity or from low dispersal rates may arise from adaptation to the same or similar regimes rather than common ancestry. Cooper et al. (2011) addressed this methodological challenge by using the Brownian rate parameter, σ^2, to measure rates of evolution across thermal niches. Theoretically, taxa characterized by high PNC should be constrained by shared conservative traits resulting in low rates of evolution. The Brownian rate parameter should differentiate among thermal niches based on variations in limiting factors (temperature, humidity, resource dispersion). Low or high σ^2 values were employed as assays for high versus low PNC, respectively. Mammals were chosen for these analyses because several large databases exist, providing requisite data.

Cooper et al. (2011) formulated four hypotheses (1) tropical clades exhibit lower PNC than temperate clades because tropical environments vary less than temperate ones; (2) taxa with small geographic ranges exhibit higher PNC than taxa with large geographic ranges because the former are exposed to a narrower range of environmental perturbations and fewer evolutionary opportunities; (3) specialists (diet, habitat, morphology) display higher PNC than generalists. The authors expected their analyses to yield low σ^2 values for tropical, small-ranged, small-bodied, and specialist mammals compared with temperate, large-ranged, large-bodied, and generalist clades with high Brownian rate parameters. Consistent with predictions, Cooper et al. (2011) found that tropical, small-ranged, and specialized species exhibited higher PNC compared to temperate, large-ranged, and generalized groups, values presumed to differentiate among thermal niches. These analyses demonstrated that results were not a primary function of body size since, considering this measure singly, small and large taxa exhibited about the same PNC. This finding may revise certain assumptions of life-history theory pertaining to the significance of body size to energy-allocation programs. It is important for mammalogists to prioritize research on the role of body size for generalizations about life-history "trajectories" relative to local and global regimes since two powerful, general statements about mammalian thermal properties based on body size metrics have recently been published (Hamilton et al. 2011; Evans et al. 2012). Resolution of these issues may depend upon the findings by Evans et al. (2012) that (1) rapid rates of trait evolution are not "sustained over long intervals in the fossil record" and (2) decreases in body size (and, presumably other morphological and/or biophysical traits) occur much more rapidly than increases, suggesting differential patterns of E gains and losses (Sect. 4.2). Theoreticians will resolve the aforementioned expansions of Evans et al. (2012), including variations in structures, mechanisms, and functions associated with "deconstraint" [Kirschner and Gerhart (1998; Chap. 5)] and evolvability of beneficial deconstraint.

Cooper et al. (2011) demonstrated that species exhibiting high PNC (narrower thermal niches) inhabited more homogeneous temporal and spatial environments, confirming their predictions that taxa characterized by low σ^2 values were more likely to be small, tropical, and specialized forms. The same study found that ecological factors (local variables) and recent evolutionary events, rather than phylogeny, were better predictors of their results. Increasingly, thermal features of microhabitats, including temperature and dispersion of limiting resources, are being recognized as important determinants of survival and reproductive success for "multicellular terrestrial organisms." Though the aforementioned study may provide insight into and stimulate further work on patterns of PNC in mammals, their proximate and ultimate causes and consequences, a recent report by Evans et al. (2012) highlights not only the availability of alternative models to calculate rates of evolution, but also strengths associated with precise specifications of the temporal features of evolution, including their constraints. The previous scientific project expresses evolutionary rates as haldanes (h), the standardized, "proportional change in a feature (M_i) between two time points" (Evans et al. 2012). Empirically based, quantitative, particularly, mathematical, treatments of evolutionary rates among mammalian taxa are in the early phases of development and interpretation.

Analyzing endotherms' functional traits, both of the aforementioned research projects highlighted the importance of assessing variances, in addition to means, when studying within- and between-individual "core processes" (temperature, variations in body mass) relative to variations in microclimate properties (air temperature, wind speed, radiation) and variations in individual traits (fur, body mass, conformation). Porter and Kearney (2009) pointed out that the global success of endotherms depends on their abilities to combine and recombine "core processes" to maintain "fixed core temperatures" across different environmental regimes. Significant metabolic costs attend the combinatorial capacities allowed by endothermy, in particular, energy and water exhaustion. According to Porter and Kearney (2009), endotherms are challenged to maintain sublethal basal metabolic rates (BMRs) within a thermoneutral ("comfort") zone minimizing energy: water (ratio) depletion. Studying Australian brush-tailed possums (*Trichosurus vulpecula*) in the field, the previous authors investigated the costs and benefits for "core processes" during animals' forays across environmental gradients, measurements circumscribing maximum time spent in TNZs (thermal niches) as a function of body mass and other characteristics associated with "core processes," and, presumably, survival and reproductive success. Porter and Kearney (2009) concluded that such analyses are "critical" for an understanding of "biophysical ecology," thermal niches, and thermal tolerances of endotherms.

Endotherms (mammals, birds, and a few other taxa) regulate and balance heat exchange between body and external environment, maintaining input and output above levels lethal to the organism. The physiological mechanisms regulating heat transfer are energetically expensive to maintain, imposing constraints on biological systems whose homeostatic processes depend upon availability of limiting resources (food, mates, breeding sites). Continuing to follow McNab (1974), Porter and Kearney (2009), Cooper et al. (2011), "decisions" made by endotherms are not only a function of exogenous variables such as resource dispersion, type, and quality, but also of other biotic and abiotic factors (temperature, intraspecific and interspecific interaction rates, body size). These and other studies have shown that BMR and minimum "thermoregulatory" limit (the "thermoneutral zone": an individual's lowest nonlethal threshold of "thermal conductance") increases with body size and absolute energy demands and that large body size is associated with decreasing heat loss per unit mass as well as decreasing neuromuscular costs (locomotion, behavior). Although patterns are not straightforward, in general, higher BMR and lower heat exchange ("thermal conductance") obtain for temperate and arctic endotherms, while opposite patterns characterize tropical species.

These and other factors varying with thermal niches (developmental rates, reproductive investment) have eluded quantitative modeling, not only because the factors change over space and time, but also because their interrelatedness and autocorrelations are dynamic, often dependent upon age and sex composition of populations, as well as mean relatedness of individuals, and other factors (resource dispersion, type, and quality). Individual reaction norms reflect direct and indirect interests, nonlinear factors potentially constraining population fitness and sensitivity to environmental change. I suggest that kin selection will often oppose group interests, particularly where competition between groups is more intense than competition within groups since the latter condition is expected to favor strong nepotistic effects. Further, since plastic responses are likely to increase interindividual differences, ceteris paribus, polymorphisms and polyphenisms may not be favored under conditions of kin–kin favoritism, possibly explaining some cases of niche conservatism and genetic homogeneity.

Terblanche et al. (2011) demonstrated the difficulty of obtaining requisite measurements to test micro-climate variables, including their spatiotemporal effects on individuals and populations. These authors discuss how to measure thermal tolerances, emphasizing the importance of assessing rates of temperature change. In particular, Terblanche et al. (2011) pointed out that field and laboratory experiments have measured animal responses to sudden rather than gradual shifts in temperature, compromising the reliability of statements about causes and effects of these phenomena for biological organisms. Studying variations in thermal characteristics within and between British habitats ("patch"), Suggitt et al. (2011) found similar or greater local effects compared with global (climate) heterogeneity. Consistent with expectation, open habitats exhibited higher temperature variability than canopy forests for most months. Within-habitat differences were explained primarily by leaf and cloud cover, topography (slope), and dispersion of vegetation. Between habitats, however, elevation was a more powerful predictor of differences in temperature. Discussing the implications of their results for thermal "grain," the aforementioned researchers emphasized the importance of determining what traits allow organisms to escape potentially lethal temporal and spatial effects.

Suggitt et al. (2011) concluded that, ceteris paribus, global ("conventional") modeling approaches are likely to permit sufficiently sensitive estimates of environmental variability excepting those species intolerant of extreme thermal parameters. These authors call for a revised conceptualization of "habitat" incorporating thermal tolerances and local "associations" of different species. Their results suggest that, even for "flexible" organisms, small shifts in one or another thermal parameter may have large nonadditive effects. It is, thus, important to consider potential consequences of shifts in microclimates for populations. Studying a terrestrial ectotherm (spider), Suggitt et al. (2011) found support for their propositions. However, such taxa are most likely to experience environments as coarse-grained relative to their generation times, conditions theoretically expected to favor robust over plastic traits (Box 2.1).

The studies reviewed so far in this section highlight challenges associated with testing hypotheses about differential effects of micro- and macroclimate heterogeneity. Whatever the abiotic and biotic factors of interest, research programs led by

basic scientists interested in questions specific to robustness, plasticity, and evolvability may not have an applied focus even though many conservation biologists
are applying sophisticated quantitative methods to study the consequences of climate change for population parameters. Reed et al. (2010), for example, analyzed their
large database of developmental, physiological, behavioral, and reproductive information on individuals from a broad range of plant and animal species, including
mammals, demonstrating alterations in population genetics and dynamics, phenotypes, and life-history parameters resulting from significant changes in thermal
regimes. Primarily interested in assessing the limits of phenotypic and genetic
responses to extreme environmental stressors (temperature, wind velocity, humidity), Reed et al. (2010) concluded that insufficient evidence exists to identify the
precise environmental stimuli constraining a population's adaptive potential upon
which effective decisions for conservation of biodiversity can be based.

Reed et al. (2010) not only clarify what variables require investigation but also
constraints on predictive models. Nonetheless, some research results permit broad
estimates, if not confident predictions, about animals' responses to increased environmental stress. Of particular interest is the development of quantitative models
predicting the realized capacities of reaction norms responding to stressors above
lethal levels. Studying Arctic polar bear (*Ursus maritimus*) populations in the
Beaufort Sea, Hunter et al. (2010) used matrix population models to forecast effects
of expected loss of ice cover on vital rates, in particular, survivorship (l_x) as a function of age and breeding status. Over a 5-year period of changing sea ice indices
correlated with climate change parameters, capture–recapture data comprising polar
bear morphometrics were subjected to bootstrapping techniques permitting researchers to project demographic fluctuations, including declining rates of population
growth. Additional quantitative methods permitted these researchers to interpolate
their results relative to historical ice cover conditions extracted from records of
greenhouse gas emissions, suggesting that basic scientists would find cooperative
arrangements with conservation biologists beneficial because of the latter's large
databases on animals and environments combined with combinatorial and analytic
tools to analyze the information.

3.2   Is Genetic Heterogeneity an Advantage in Fine-Grained Conditions?

Discussing the well-documented association between large body size and low
allozyme variation in mammals, Selander and Kaufman (1973, Nevo 2001) concluded that sufficient genetic variability is maintained in these populations as a
result of dispersal and selection. Their reanalysis of previous data included rodent,
cetacean, and human (European) samples, yielding results consistent with Levins'
(1968, in Selander and Kaufman 1973; in Emlen 1973) theory of "adaptive monomorphism" (Sect. 3.1). Shafer et al. (2012), studying North American mountain

goats (*Oreamnos americanus*), demonstrated that low immune gene diversity arose before the Pleistocene glacial "bottleneck."

Research on a diverse and widely distributed Brazilian didelphid (*Monodelphis domestica*) (Caramaschi et al. 2011) exemplifies the conceptual framework proposed by Suggitt et al. (2011). Using mtDNA samples, Caramaschi et al. (2011) compared *M. domestica* populations across "morphoclimatic domains," revealing correlated variations between opossum genetic structure and differences in thermal characteristics of habitats in addition to associations among phylogeny, climate history, and current conditions. However, Nabholz et al. (2008), studying 277 mammal species, failed to detect associations between mtDNA diversity and any of 14 life-history variables, including body mass. Similarly, mitochondrial genetic diversity was unrelated to geographic range or IUCN Red Book status, demonstrating, instead, high variability within and between taxonomic levels and suggesting that mtDNA diversity does not predict species abundance. Nabholz et al. (2008) concluded that stochasticity has been a major determinant of effective population sizes in the Class, probably explained by variations in mutation rates among mammal taxa. Nabholz et al. (2008) and Cooper et al. (2011) suggested that variations in evolutionary rates must be specified in predictive models of mammal character traits and diversity, including differential responses by individuals and populations to variations in thermal conditions.

Miller et al. (2011) tested the widely held assumption that environmental stochasticity is associated with high levels of genetic heterogeneity, implying that stressors impacting thermal tolerances place heavy demands on molecular and cellular mechanisms. These authors demonstrated that low genetic diversity in "Tasmanian devil" (*Sarcohilus harrisii*) populations preceded their current critically endangered status by at least a century. Drastic declines in populations of this carnivorous marsupial endemic to Australia are occasioned by virally induced cancer; thus, disease susceptibility has not resulted from incrementally deleterious effects of monomorphism. These authors analyzed correlations between genetic heterogeneity and IUCN Red Book status for 14 mammalian taxa. Although "Tasmanian devils" and (extinct) "Tasmanian tigers" (*Thylacinus cynocephalus*) exhibited very low genetic diversity, there was no straightforward relationship between that variable and extinction risk for the sample group as a whole, and, with the arguable exception of the two "Tasmanian" species, all of these animals are large-bodied. If mammalian mutation rates are strong effects of environmental stochasticity, variations in genetic diversity are not expected to be highly correlated with rates of evolution in mammals, since mutation rates are effects of environmental stochasticity, as indicated by Nabholz et al. (2008) and Cooper et al. (2011). This proposition is testable with available data.

3.2.1   Howler Monkeys as Assays for Studies of Genotypic and Genomic Responses to Environmental Heterogeneity

Mantled (*Alouatta palliata*) and red (*A. seniculus*) howler monkeys inhabit a very broad range of habitats and forest types (Crockett and Eisenberg 1987). Red howlers are more tolerant of xeric conditions compared to congenerics, exhibiting a significant degree of genetic heterogeneity, contradicting inferences from "adaptive monomorphism" theory. *Alouatta* is a relatively well-described and highly diverse genus, well suited to study theoretical and empirical questions related to thermal tolerance evolvability. For example, compared to congenerics, endangered black howlers (*A. pigra*) utilize a relatively narrow niche and exploit food of relatively lower quality compared to mantled and red howlers. Black howlers preferentially inhabit wet or moist forest habitats with mean annual temperature ~25°C and rainfall >1,000 per year. Forest edges are usually avoided, and the species is most often found at altitudes <1,300 ft. If black howlers are more specialized than other species in the genus, environmental stress should impose relatively greater constraints upon them, conditions thought to favor phenotypic variability (Fig. 2.2), a proposition supported by descriptive studies of this species (intentional "hand-holding" solicitous paternal care). Investigating a broad range of ecological and evolutionary questions with howler monkeys might enhance mammalogists' understanding of other folivorous mammals (sloths, some rodents, some primates, porcupines, armadillos, some marsupials, some bats, moles, colugos), in addition to general questions related to evolution in heterogeneous regimes.

Howler monkeys fill an arboreal niche sometimes compared to terrestrial ungulate niches, potentially providing comparative tests of propositions concerning genetic variation, convergent evolution, and energy-allocation strategies. Howlers prefer a diet of new leaves, flowers, and fruit but are also capable of exploiting more evenly distributed mature leaves (Belovsky 1984). Red howlers (*A. seniculus*) exhibit a broad geographical range that has not contracted in response to anthropogenic events, and these primates exhibit significant genetic heterogeneity, possibly "buffering" individuals from effects of potentially lethal environmental stressors (Crockett and Eisenberg 1987). Black howlers (*A. pigra*), on the other hand, are characterized by high levels of genetic monomorphism (James et al. 1997) and a geographic range limited to Belize, northern Guatemala, and southeastern Mexico, while mantled howlers (*A. palliata*), the basal taxon of the genus (Villalobos et al. 2004), display the widest geographic distribution of any howler species and, according to one highly publicized dissertation, low levels of allozyme variation (Malmgren 1979). Since *A. palliata* is the basal taxon of its genus, its genetic profile is probably the conserved state. Independent of particular patterns of genetic diversity characterizing howler species, however, the widely distributed genus is well suited to address aforementioned topics as well as others of import for all vertebrates, such as variations in genetic traits along abiotic and biotic gradients. For this and similar research programs, Rutherford's (2000) proposition applies: "Genetic variation is pervasive, but its expression as phenotypic variation is context-dependent and heavily buffered."

Jones (1997b) summarized literature on "rarity" in primates, finding three generalizations (1) Abundance may be correlated with extinction if populations with low densities (α-rarity) are at or near levels of negative growth (competitive inferiority, as well as sensitivity to perturbations and changes in carrying capacity). Even where their abundances are low, insectivore-carnivores appear particularly resistant to environmental perturbations; (2) Habitat specificity may be "broad" or "restricted" (sensitive to spatial "patterning" and size), and species with a restricted habitat range (β-rarity) may be "extinction prone." Few mammals are "habitat specific" in the sense that many insects or plants are. Indeed, many mammals are labeled "broad habitat specialists" (Box 2.1, Sect. 3.3). Habitat specificity and edaphic categorization are often correlated, and estimates of extinction risk should measure (thermal) niches (for a "quick and dirty" method: Hanya et al. 2007). Frugivore-herbivores appear to be particularly resistant to potential risks associated with β-rarity; (3) Probabilities of extinction are area dependent, and the geographical distribution of a species may be "wide" or "narrow" (γ-rarity). Genera with the largest geographical range are not necessarily the most speciose (*Cebus* compared with *Alouatta* among Neotropical monkeys), and widely distributed genera heavily dependent on fruit (*Cebus*) may reflect a fine balance of "plant–animal interrelationships" with the potential, when shocked, to effect population "crashes" or "faunal collapse." Fleming et al. (1987) identified different spatiotemporal patterns of fruit availability when Paleotropical (Asia and Africa) and Neotropical (Latin America) forests were compared, effects requiring evaluation in research on extinction since fruit is commonly a dietary component of abundant species. Cody (1986 in Jones 1997b) posited that "γ-rarities" are most vulnerable to extinction because they are usually endemics, although these same species are often locally abundant. The same author concluded that "γ-rarity" is most likely to occur in taxa with the greatest potential for isolation (islands) and diversification (cf. folivorous Cheirogaleidae, Lemuridae, Indriidae) and that stochastic processes (islands) as well as historical factors probably explain these results rather than deterministic ones (community effects).

3.3   Spatiotemporal Variability in Microclimates and Macroclimates: Seasonal Forests as Natural Laboratories for Theoretical and Experimental Research on Mammalian Thermal Tolerances

Flexible mammalian adaptations (endothermy, big brains, broad thermal tolerances, polygynandry) were driven by the rigors of survival and reproduction in heterogeneous regimes (Fig. 2.2), in particular, shifting resource dispersion, type, quality, and the like (food, mating sites, mates). First principles of ecology require that the size and composition of mammalian population structure, particularly, density and sex ratio relative to characteristics of food supply, change in response to temporal environmental heterogeneity with important consequences for the survival and fecundity of individuals (Eisenberg 1981; Jones 2009). Mean population abundance and structure over time, including group size, is an attribute of resource predictability. Ceteris paribus, high resource predictability and high resource quality, relatively homogeneous spatial dispersion of resources and resource tracking by the animal population is expected to favor resource defense (contest competition, territoriality) by individuals or small groups, whereas low resource predictability and large distance or high variation in distance between patches may make resources indefensible (not monopolizable), yielding large mean group size. Since temporal unpredictability of resources may be positively correlated with spatial uncertainty ("patchiness"), foraging in groups may reduce mean searching time per individual group member. Thus, ceteris paribus, environmental predictability will be inversely correlated with group size and composition, although this ratio may vary over time and space.

Apparently, the benefits of group life have not outweighed sociality's costs for males in most mammalian taxa since sexual segregation of adults is characteristic of the Class. Nonetheless, complex social behavior in which adults of one or both sex exhibit tolerance and consequent spatiotemporal integration may be found in several mammalian orders (carnivores, cetaceans, rodents, primates, bats). Some social mammals are eusocial or breed cooperatively, exhibiting noteworthy physiological and behavioral mechanisms of reproductive and social control, repression, cooperation, and, in Bathyergidae, primitive eusociality. For group-living mammals, reproduction is often a selfish and costly act, requiring trade-offs between allocation of energy to mating (usually males) or parenting (usually females), and to the consequences of these decisions for the breeder, her offspring, and the individuals comprising her social space (especially kin). Decisions to breed, in addition, may increase competition with other reproductive females for existing resources, especially food, sexual partners, and space (breeding sites).

Since parameters describing norms of reaction and relative reproductive success of females in populations are important determinants of a population's life table and mean fitness, it is important to unpack the regulatory feedback mechanisms, trajectories, and functions controlling strategic conversion of energy into offspring, responses sensitive to endogenous and exogenous sources of information (particularly, temperature: Meyer et al. 2011; Heidel et al. 2011). Research on invertebrates and vertebrates suggests strongly that females are the more canalized or buffered sex and that, relative to conspecific males in similar conditions, female phenotypes, ceteris paribus, are less responsive to environmental fluctuations compared to males in comparable conditions (Lerner 1954; Agrawal 2010; Jones 2005a; Box 2.1) because of metabolic constraints. The latter observation leads to the prediction that, relative to males in the same population, adult females are more likely to display variable energy-saving and/or energy-compensatory tactics and strategies. As noted in Sect. 4.1, phenotypic variability may characterize females; however, relative to males in comparable conditions, plasticity is expected to stress females' "fitness budgets", demanding a proportionately greater investment of "free energy" to reproductive-allocation strategies, imposing relatively higher ecological stringency.

Fig. 3.2 Worldwide distribution map, tropical seasonal forests. In the present review, seasonal forests and other heterogeneous regimes (tundra, desert) are recommended as laboratories for within and between taxon studies of differential patterns of reaction norm, life history evolution, and evolvability in mammals (Chaps. 3 and 4)

In order to test these and related ideas, it is necessary to specify the endogenous and exogenous structures, mechanisms, and functions associated with differential tolerances × sex, and, most likely, other factors (age), to micro- and macroclimates (Sect. 3.3).

Spatiotemporal heterogeneity induces stress responses in biological systems, making seasonal forests (Fig. 3.2) a good laboratory for scientific investigations of challenges associated with survival and reproduction, from the organism's molecular and cellular to higher levels of organization (Table 3.1). Although plants and animals experience disadvantages associated with inhabiting fluctuating environments, a major advantage is that time-varying patterns associated with seasonality yield environmental predictability, providing cues and signals about changing conditions (e.g., changes in sunlight, humidity, or rainfall patterns). The discrete stimuli signaling wet and dry seasons in time-varying environments modify physiological and behavioral processes of organisms, allowing adjustments to the stress of different temporal features (e.g., abundance and distribution of nutrient and water resources).

Ecologists studying plants and animals inhabiting seasonal forests measure patterns of mortality to evaluate the success of organisms' responses to temporally varying regimes (Stearns 1992; Hallgrímsson and Hall 2005). These studies demonstrate "trade-offs" between adult mortality and mortality of immature age-sex classes, and mammalian adult survivorship is commonly favored in seasonal forests. Ecological studies have shown that this life-history trade-off results from uncertainty and risk associated with reproducing in fluctuating regimes in which, compared to wet tropical forests, cues and signals predicting the initiation and termination of seasons are highly variable. These challenges may stress regulatory, developmental, physiological, and behavioral processes influencing lifetime reproductive success of individuals in a population. Statistical analyses of time-varying features are predictable on average, but the onset, duration, and termination of temporal patterns within and between

Table 3.1 This source table (ANOVA) of chest circumference (CC: dependent variable) × habitat (independent variable) for adult female Costa Rican howler monkeys shows a significant between-habitat difference, possibly resulting from differential heritable or nonheritable accommodation to different microclimates (Jones 2006)

Source	SS	Df	MS	F	P
Habitat[a]	1,318.46	2	659.23	3.5986	0.0304
Residual	21,433.5068	117	183.1921		
Total	22,751.9667	119	191.1930		

CC is significantly smaller for females in the most degraded, irrigation habitat relative to female CC for females in riparian or deciduous habitats (irrigation < riparian, deciduous; riparian = deciduous). Comparable analyses for adult males using ANOVA yielded no significant results. The significant comparisons for females indicate the operation of differential allocation of energy to cardiovascular function(s) as reported for children in India (Sundaram et al. 1995 in Jones 2006). The present study and the study in India, strongly suggest that morphometric-environment concordances are explained by epigenetic effects in response to differential environmental factors such as overall patch quality, temperature, nutritional variables, and food availability. In irrigation habitat, a statistically significant relationship ($r=-0.3895$, $p=0.005$, $n=39$) was found between CC and pubis width for the same sample of adult females in the three habitat regimes, suggesting the possibility of an energetic trade-off for females in the most degraded thermal niche (habitat) as well as a unique thermal tolerance profile. It is important to explicate that several of the significant differences found for both adult males and females using correlation statistics (Jones and Agoramoorthy 2003) were not confirmed after treatment with ANOVA (Jones 2006). This methodological inconsistency may have implications for research projects addressing similar questions with other mammals, as well as vertebrates, in general. Additional details, analyses, and extensive discussion of habitat–morphology associations in mammals are available in Jones (2006).
[a]Irrigation × Riparian (♀♀): $p<0.05$ (Newman–Keuls test); Irrigation × Deciduous (♀♀): $p<0.05$ (Newman–Keuls test)

years are usually difficult for organisms' sensory and perceptual mechanisms to assess. Ambiguity and error resulting from between-cycle variability usually leads to significant mortality in one or more age-sex class in a population (Stearns 1992).

Following several authors (Stearns 1992; Eisenberg 1981; Jones 1997a; Allee 1931; Vaughan et al. 2000; Emlen 1973; Box 2.1), mammals and other organisms inhabiting seasonal tropical environments may respond to perturbations with one of several life-history strategies, each of which entails both costs and benefits. Organisms may display *inter- or intraspecific parasitism* (Jones 2005a, b, c; Discussion), often exemplified, in mammals, by *kleptoparasitism* (one species parasitizing another's food resources). The parasitic species (sea lions parasitizing dolphins' foraging strategy) depletes some proportion of the host's energetic "fitness budget." Parasite–host associations are usually specialized with respect to the species involved, and measurable (proximate and ultimate) costs to hosts (dolphins) represent either a reproductive dead end or a condition whereby mean reproductive benefits outweigh costs, relative to other strategies available to hosts.

Another reproductive pattern entails *specialization to the most common forest type* (*microclimate*), such as riparian or deciduous habitat of tropical dry forest environment. Specialists are usually characterized by a diet limited in food types and a tolerance for a narrow range of environmental conditions (pandas). Mammals are not usually habitat specialists, but food specialization is not uncommon (nectar-feeding bats, woolly flying squirrels, and pine needles). In such cases, distribution of animals in habitats will be constrained by food dispersion and quality, and, for many, if not most, mammal taxa specialization is more apparent than real since cryptic food tolerances may be expressed under conditions of food restriction (armadillos), a common pattern of resource use among folivorous mammals, compensating losses with relatively even, continuously distributed foliage, buffering populations from environmental unpredictability. A specialist strategy may be beneficial when conditions favor specialized traits; however, specialists may be highly vulnerable if environmental regimes change significantly.

Many mammals, especially, large ones, are habitat and food *generalists*, exhibiting a "compromise" or a "mixed" strategy across conditions (pika, mantled howler monkey, Pallas's cat; Box 2.1, Sect. 3.2). Some generalist species exhibit a phenotypic compromise or "mix" of traits advantageous to an intermediate "fit" across changing conditions (across both deciduous and riparian habitats in Central American tropical dry forest environments), and the optimal generalist foraging strategy is highly sensitive to energy intake per unit of foraging time. Generalist phenotypes are the most common ones displayed by plants and animals in tropical seasonal environments and, all other things being equal, generalists thrive in a wide range of thermal conditions, utilizing a variety of dietary sources. Intuitively, compared to other life-history modes, generalist strategies may seem more protective and least costly to lifetime reproductive success. However, generalists may be "spread too thin" (over time and space) to respond successfully to extreme environmental perturbations (climate change, habitat destruction) and, importantly, generalist life histories may be severely constrained by trade-offs between foraging requirements and predator risks (variations in cover). Mayr (1976) noted that, across animals, generalists are "widespread and abundant," and the most well-described and geographically successful generalist mammal is the omnivorous, *Homo sapiens*.

Other mammals inhabiting seasonal habitats *specialize on one of the environment's morphs* (dry or wet; cold or warm). A number of these taxa are migrating (wildebeest) or hibernating (black bears; torpor: echidna) species, exhibiting behavioral thermoregulation (Hanya et al. 2007), "escaping" negative effects of the least favorable suite of conditions. This life-history strategy is common in birds, but not in mammals, possibly due to the efficient combination of endothermy, relatively generalized phenotypes, and broad ecological tolerances among members of the latter Class. Among species in this category (Tundra wolves), "hierarchical" habitat preference ("prioritization"), rather than specialization to one habitat, is the norm, and these mammals are noteworthy for their dispersal and colonization abilities highly sensitive to spatial scale. Mayr (1976) concluded that, across animals, specialization (a derived thermal mode) to a narrow niche is "perhaps the most common evolutionary trend," particularly, among insects, a feature that this author considered an effect of the order's "prodigious rate of speciation," ultimately caused by ecological constraints.

Mammals commonly display *seasonal breeding* (in plants, "mast fruiting": *Manilkara* spp.), whereby reproduction is synchronized across females in varying degrees of temporal overlap (short-beaked echidna). Apparently, this allocation strategy is favored to mitigate preweaning mortality, is associated with high confidence of fertilization for males, and is likely to be observed where females mate with a single male per breeding season. *Biannual reproduction*, another adaptation to seasonal environments observed in some mammals, is thought to represent a "lifetime energy-conservation strategy" in regimes characterized by very high variability in food availability (differential abundance), dispersion (evenness), quality (toxicity), and the like (canids, rodents, bats, flying squirrels). Biannual reproduction increases predictability of and information available in local ("patch," microclimate) conditions, benefits trading off with small litters (relative to body mass and foraging strategy), as well as relatively short periods of gestation and lactation. In critically stressful conditions (near-lethal thermoregulatory stress), biannual reproduction may grade into an *opportunistic* (facultative, "random") pattern observed in some small, tropical rodents (wild guinea pig). The young of opportunistic (aseasonal) breeders are precocial, while seasonal breeders produce altricial offspring.

3.3.1 Atelidae as Protocols for Research on Mammals in Heterogeneous Regimes

For each of the aforementioned life-history modes, the evolutionary and ecological strategies characteristic of specific populations depend upon a suite of factors, including ancestral and derived traits as well as differential life-history metrics describing populations [generation time (Box 2.1; Sect. 3.2; Jones 1997a), rate of intrinsic increase (r), and net reproductive rate (R_0)]. In the present review, howler monkeys (*Alouatta*) are advanced as exemplars of a broad range of questions related to robustness, plasticity, and evolvability as well as regulatory feedback mechanisms underlying the three states. Several rationales recommend howlers and other atelids as models for tests of most questions related to effects of variations in exogenous (abiotic and biotic) agents interacting with phenotypes of primary consumers and their endogenous properties (molecules, cells, genomes, neurotransmitter substances, physical characteristics). The five genera in the primate family, *Atelidae*, are howler, *Alouatta*; spider, *Ateles*; wooly spider, *Brachyteles*; wooly, *Lagothrix*; and yellow-tailed wooly, *Oreonax*.

Atelids are among the most thoroughly documented social mammals, exhibiting widely diverse occupation of forest types and habitats within and between species; significant variability in life histories and population biology, narrow (*Ateles*, *Brachyteles*: derived) to broad (*Alouatta*: conserved) thermal niches, including some combination of the following traits: dietary composition; a range of mid-sized body masses; conserved, generalized phenotypes in combination with a variety of stereotyped motor patterns and displays permitting noteworthy behavioral flexibility; robust and variable reproductive parameters (iteroparity; life-history strategy) favoring female survivorship; "stairstep" survivorship function characteristic of herbivores; "bet-hedging" reproductive tactics; low variability of gestation time among congenerics;

weaning period truncated, less variable, compared to many primates; relatively low intrinsic rates of increase compared to congenerics; limited parental investment; male philopatry or bisexual dispersal sensitive to habitat quality; colonization; "targeting" as a female aggressive tactic, maybe a precondition for female dispersal and/or colonization; characteristic variable rates of infant and juvenile male survival; differential IUCN status; energetically efficient behavioral tactics and strategies, including locomotion and "rule-governed" foraging, possibly related to costs of folivory; associations among unrelated individuals; similar to humans, reliance upon vocal, rather than tactile, communication, a derived trait; aggressive restraint, suggesting high social grades.

Morphometric data × habitat are available for *A. palliata* (Table 3.1) permitting preliminary assessment of variations in phenotypic traits relative to variations in ecological gradients. Ultimately, lifetime reproductive success is a direct function of reaction norms (Environment × Genotype interactions and the "responsive phenotype"), and data permitting measurements of relative reproductive success of females × habitat are available for some atelids (*A. palliata*; Fig. 3.3). From the molecular to the social, population, and community levels of organization, including their status as sister taxon to congenerics combined with relatively intermediate symptoms, recommend mantled howlers for intense study as a typological model not only for the genus, *Alouatta*, but also for the class as a whole. Sequencing of the *A. palliata* genome would complement nucleotide databases for other mammals (platypus, red kangaroo, macaque), permitting fundamental comparative analyses at the molecular level and a quantitative foundation for understanding the success of many atelid species under changing conditions and population decline. From the female's point of view, there remains the problem of optimal timing of reproductive allocation across the life span to reduce effects of environmental perturbations and to maximize likelihoods of her access to, monopolization and processing of limiting resources (E, Q), as well as of her own survival in the service of future reproduction (Stearns 1992).

3.4   Thermal Tolerances of Mantled Howler Monkeys Are Preadapted to Stressful Heterogeneous Regimes

Chevin et al. (2010) proposed "mechanistic approaches" for analyses of "ecological and evolutionary responses" ("environmental [thermal] tolerance curves") to climate change (Box 3.1; Schoener 2011), applicable to many of the examples in the present brief. Support for the utility of mantled howlers as a model for the investigation of questions related to thermal niches in mammals is indicated, in part, by the taxon's mosaic phenotypes (independently evolving traits) characterized by conserved and derived, including ritualized, characteristics in combination with noteworthy behavioral flexibility. These wholly herbivorous, diurnal primates inhabit seasonal tropical forests (riparian and deciduous habitats: Jones 1997a) from southern Mexico to northern Ecuador. Reanalyzing Glander's (1975) data on foraging mantled howlers

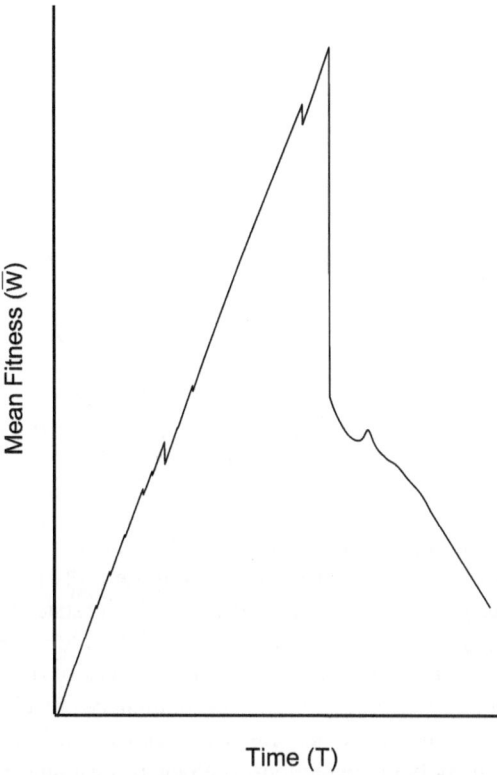

Fig. 3.3 A simple drawing illustrating population increase toward realized mean fitness (*peak*) and subsequent decline, often a "downward spiraling cascade" with potential for population and other network (interspecific associations) collapse. Extinction ("point of fracture"), a deterministic process whereby population viability is threatened by (usually) rapid decline in numbers, may occur wherever environmental fluctuations (heterogeneity, unpredictability) expose members of a population to stressors beyond their capacities to respond (system insult and trauma, near-lethal conditions). In these regimes, mortality may outweigh reproduction, eroding population size and structure, decreasing intrinsic rates of increase. There is wide agreement among evolutionary ecologists that a population's reservoir of genotypic and phenotypic variation is usually sufficient to compensate for increased environmental fluctuations (Nevo 2001). However, some perturbations are so severe ("knockdowns") that population recovery and replacement are obviated. Gomulkiewicz and Holt (1995) emphasized the importance of characterizing "… those combinations of genetic and demographic conditions likely to result in persistence versus those expected to lead to extinction in a changed environment." With theoretical constructs, the previous authors explored selective processes capable of stalling or reversing decline of a population's intrinsic rate of increase, concluding that a population's genetic potential for "rapid evolution" is critical for its recovery from declining numbers

at Hacienda La Pacifica, Cañas, Guanacaste, Costa Rica, Jones (1996a) demonstrated significant seasonal variations (temporal heterogeneity) in availability of food utilized by the monkeys. Theory predicts that animals will "track" ("demographic race") their environment with behavioral and physiological rather than genetic mechanisms

Box 3.1

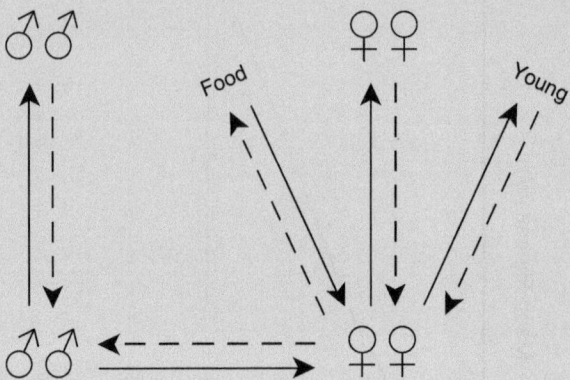

Suggested directions of potential conflicts (differential optima) where individuals of one class or category of organisms impose costs in inclusive fitness upon individuals of another class or category (initiating state, closed arrows) to which the latter may respond (broken arrows). These coevolutionary states may manifest as mutualism, parasitism, or other types of (recurrent) interactions (exploitation). Assuming that the coevolutionary processes do not slow down or terminate (one individual effectively represses the other, termination is beneficial to one or both interactions, ecological constraints obviate an optimal response), interactants may be, in one manner or another, recurrently reprogrammed over (generational) time. The scenarios depicted in this image incorporate two different routes whereby coevolution within populations is initiated and, presumably, maintained. In the first scenario, coevolution between reproductive females and plants is the primitive condition from which other conflicts leading to coevolutionary interactions emerge (The Emlen–Oring Scenario). In the second scenario, male parasitism of females is the primitive condition driving the other patterns of exploitation (that may be beneficial to receiver as well as sender). Dissection of these scenarios, including, their evolutionary consequences and likely nonindependence, is beyond the scope of the present review. However, it is worth highlighting that all dyadic interactions, including animal–food interactions, are amenable to analysis with quantitative models of parasitism (Jones 2005c; Discussion).

Mammalogists have probably not emphasized the role of intraspecific parasitism as a possible determinant of population structure because the pertinent models have been associated with the insect literature. Any of the sender–receiver interactions displayed in the present figure may be characterized as a parasitic "relationship" between organisms having different allocation optima and, by definition, a relationship in which the parasite is to some degree

(continued)

Box 3.1 (continued)

dependent upon its host. The present propositions ignore possibilities for parasite- or host-switching, forms of phenotypic variability very likely to apply to many mammals, in part, because of their highly differentiated nervous systems (though this biological feature is not a necessary condition for switching hosts or parasites or for switching from one "role" to another). All types of intra- and inter-organism parasitism concern access to limiting resources regulated by positive and negative feedback processes, including systemically controlled mechanisms designed by evolution to maximize reproductive success (Poulin 2002; Jones 2005c). Among endotherms, interspecific parasitisms occur in association with breeding ("brood parasitism": birds) and feeding ("kleptoparasitism": African wild dog, *Lycaon pictus*; spotted hyena, *Crocuta crocuta*) with intraspecific analogues (manipulation, harassment, exploitation, dependency). As formulated in the Discussion of this review, these economic (energy allocation) relationships are amenable to straightforward quantitative modeling. ©Clara B. Jones

where environmental changes are briefer than generation time (Slobodkin and Rapoport 1974; Box 2.1), calculated as 6.27 years for howlers at this study site (Jones 1997a). Time-series analyses of rainfall in Guanacaste revealed strong signals at ~6 months cycles (seasonality, Box 2.1), a predictable pattern briefer than T, creating a "fine-grained" environment for these anthropoids (Jones 1997a).

Glander (1975) showed that his study group spent ~24% of their yearly activity "budget" feeding. Six plant families accounted for ~75% of howler feeding time, three of these (Anacardiacaea, Mimosacaea, Papilionaceae) accounting for ~61% of total feeding time with ~18% of the total spent feeding on flowers, including buds. Glander (1981) demonstrated a high degree of prey selectivity by howlers, preferring flowers, leaf flush, and fruit of five tree species (*Albizzia adinocephala, Enterolobium cyclocarpum, Inga vera* var. *spuria, Pithecolobium longifolium*, and *P. saman*). The previous five species are among the 25 species used most often for food by howlers with discriminative feeding occurring in response to phenological patterns within and between seasons, habitats, species, and individual trees characterized by qualitative and quantitative differences among plant parts over time and space.

Glander (1975, Table 41) presented data for the number of species out of 25 producing new leaves, flowers, or fruit each month. Table 3.2 presents these results. In absolute terms, there are more species producing new leaves than flowers and flowers than fruit in each month except June, July, and August (wet season) during which more species are producing fruit than flowers. These calculations do not take into consideration variations in sizes of trees, fluctuations in phenophases within and between months, or hierarchical food preferences that might govern temporal and spatial dispersions of the monkeys. Table 3.2 permits relative assessment of preferred plant tissue availability per month.

Table 3.2 Number of species producing new leaves, flowers, and fruit per month for the 25 tree species preferred by mantled howler monkeys at Hacienda La Pacifica, Cañas, Guanacaste, Costa Rica

Month	New leaves (NL)	Flowers (Fl)	Fruit (Fr)
January	10	6	5
February	16	13	7
March	17	13	6
April	17	11	7
May	18	8	8
June	11	4	7
July	11	4	5
August	11	2	4
September	13	5	4
October	15	4	4
November	11	5	2
December	12	5	4

See text for explanation

Table 3.3 Relative reproductive success (RSS: Mean J+I/Females/Group) as a function of group size at two Central American sites where two mantled howler monkey subspecies are located: *Alouatta palliata palliata*: Guanacaste, Costa Rica (GTE: tropical dry forest, $n=51$ Groups) and *A. p. aequatorialis*: Barro Colorado Island, Panama (BCI: lowland tropical moist forest, $n=73$ Groups)

Females/Group (n)	GTE		BCI	
	RRS	F	RRS	F
2	0.75	2	0.66	3
3	0.67	1	0.17	2
4	1.00	2	1.21	7
5	0.81	4	1.23	7
6	0.99	4	1.20	11
7	0.79	6	0.99	10
8	0.55	10	1.01	15
9	0.58	8	1.03	8
10	0.55	2	0.80	4
11	0.64	1	0.82	1
12	0.75	1	1.00	2
13	0.86	5	0.66	2
14	1.00	2	1.14	1
15	0.57	2	–	–

Frequency (f) = number of times a female group of a given size (n) occurred at GTE and BCI. Modal female group size = 8 for both sites. The frequency distribution of female groups was compared between sites and the mean ($\pm$ S.D.) number of females per group was significantly larger in GTE (8.38 ± 3.24) than at BCI (7.10 ± 2.58), Randomization Test: $T=2.58$, df$=121$, $p\leq0.01$, a result that might be accounted for by the higher degree of seasonality, patchiness, and resource clumping in tropical dry forests, although both sites are characterized by relatively moderate levels of primary productivity. Howler populations, thus, appear to be limited by environmental potential (thermal niche, *cum* variability in limiting resources), with greater potential for larger group sizes in the more heterogeneous forest type (GTE). See Jones (1996a) for additional information on this analysis, showing, for example, that female productivity at each site is, statistically, equal. As suggested in the text of this review, within- and between-species comparative studies using the widely distributed genus, *Alouatta* (variations in genetics and genomics, morphology, physiology and development, behavior, social organization, ecology, geographic distribution, as well as other traits), would be instructive for mammals in general and herbivorous mammals, in particular, because of the noteworthy diversity among howler monkeys and other atelids

On average, 13.5 ± 2.9 (mean + standard deviation) of the top 25 tree species were in new-leaf phenophase per month, per year, 6.67 ± 3.73 species flowered, and 5.67 ± 1.3 species bore fruit. Coefficients of dispersion for these phenophases are 0.62, 2.09, and 0.30, respectively. Mean diversity of new leaves was greater across months than the means for flowers and fruit since more species exhibiting new leaf flush were available each month. Howlers were hypothesized to map their distribution onto conditionally available, local ("patch") dispersion of their plant prey (energy) in addition to the absolute number of individual trees per species, tree size, variations in food quality, and forest architecture (Table 3.3). All of these features are amenable to change by abiotic (weather, tree-fall) and biotic (intra- and interspecific competition, pollinator density) processes, among other factors (age, sex, and dominance-rank of howlers).

Coefficients of dispersion for new leaves and fruit are dispersed (more observations than expected around a central tendency), whereas flower dispersion is clumped (more observations than expected at tails of distribution). What do these distributions imply for howlers foraging for prey items? Clumping of species in flower shows that more of the 25 preferred species were flowering or not flowering than one would expect if flowering by different tree species were independent. This finding is consistent with results from other reports on phenological patterns in tropical forests showing flowering synchrony within and between tree species, suggesting that similar environmental signals trigger flowering at about the same time across months. Frankie et al. (1974 in Jones 1997a), for example, demonstrated that cessation of rainfall provides proximate signals inducing physiological changes triggering flower phenophases.

Repulsed new leaf and fruit distributions are more difficult to interpret than the clumped distribution shown for flowers. In statistics, repulsion does imply that new leaves and fruit are more evenly dispersed across months since about the same number of species exhibit new leaves or fruit over time. Although the clumped distribution of flowers implies that they are a highly predictable resource, they are also temporally ephemeral since duration of flowering varies within and, to a greater degree, between taxa. Among the additional challenges confronted by foraging *A. palliata* would be variations in food quality, including toxicity, within and between species as well as fine-grained differences in (gradual) onset and (gradual) termination of the flowering phase. Thus, flowers are more predictable when assessed on a global (population) compared to a local ("patch") scale. It was not possible to assess relative spatiotemporal predictability of new leaves and fruit with Glander's (1975) dataset. However, it seems reasonable to assume that conditional variations in plant food dispersion and quality compromise spatiotemporal predictability of resources, increasing the proportion of a howler's energy-budget used for acquisition, ingestion, and digestion of plant prey.

Howlers are able to extract water from their diet and are capable of utilizing mature leaves of relatively lower nutritional value than other plant items. However, both of these energy-conservation options decrease metabolic (energetic) efficiency because of physiological costs associated with populations of enzymes and microbes for extraction of moisture and energy as well as for detoxification secondary compounds in plant tissues, particularly, mature leaves (Glander 1978). Studying rodents under controlled laboratory conditions, Krause et al. (2011; Dell et al. 2011) demonstrated that thermal stress induced by dehydration inhibits protein regulatory mechanisms controlling compensatory physiological and motor responses. On the other hand, as demonstrated for several organisms (giant pandas: *Ailuropoda mela-noleuca*: Zhu et al. 2011), coevolutionary benefits from gut microbiome–host associations include enhanced metabolic homeostasis and efficiency (de Jong 1976) resulting from increased digestibility of diets high in fiber and toxins. Inferences from the present reanalysis of data displayed in Table 3.2 support the idea that similar regulatory "toolkits" are identifiable across taxa for similar functions (social traits: Bell and Robinson 2011; Keller 1995).

A coefficient of variation (CV) was calculated for the number of months preferred food was available for the top 25 tree species utilized (Table 3.2). New leaves, flowers, and fruit displayed CVs of 0.30 (8.6 ± 2.6 months), 0.87 (5.3 ± 4.6), and 0.47 (6.3 ± 2.98), respectively. Availability of flowers fluctuated more, and new leaves and fruit, less, per month, consistent with the aforementioned analysis of monthly species diversity and plant tissue type. Glander's (1975) Table 41 also permitted an evaluation of the relative degree of temporal clumping or randomness of the top 25. A "runs test" (Siegel 1956 in Jones 1996a) was performed on availability of each phenophase for the top 25. Sixteen "runs" (8 NL, 6 Fl, 5 Fr) for 14 species could not be evaluated due to insufficient frequency of "runs." Twenty-five "runs" (9 NL, 11 Fl, 5 Fr) for 16 species exhibited a random pattern of tissue availability for one or more of the three phenophases. Ten "runs" (1 NL, 5 Fl, 4 Fr) of nine of the top 25 exhibited significant temporal clumping ($p < 0.05$). These results indicate that uncertainty and resource clumping are constant components of local conditions in which howlers work to survive and reproduce. From the perspective of organisms, spatiotemporal clumping and unpredictability of limiting resources, as well as fine-grained conditions (generation time relative to periodicity of resource availability: Box 2.1; Emlen 1973; Templeton and Rothman 1974; Strobeck 1975), favors metapopulation structure, generalist phenotypes, and flexible responses to changing local ("patch," microhabitat) conditions (Emlen and Oring 1977; Fusco and Minelli 2010; Hanya et al. 2007).

Spatiotemporal variations in energy availability do not necessarily provide information about how mammals manage potentially lethal stressors or whether responses to these conditions represent cryptic, opportunistic plasticity or preadaptations, including novel responses (social learning, imitation) to uncertainty and risk (dusky dolphins, bottlenose dolphin, polar bears, Tasmanian devil, humans, Japanese macaques, killer whales, nectar-feeding bats, red squirrel). Reed et al. (2011) specified a large suite of variables, including microhabitat cues and signals, considered necessary for quantitative assessments of interacting phenotypic, genotypic, and climate (thermal) factors. As these authors point out, empirical data required to define the parameters of their integrated (verbal) model are not available for any species. It is unlikely that an "understanding" of resilience and persistence is attainable for all but a very few species within an acceptable time period advocating for use of bootstrap and other modeling approaches.

The previous reanalysis of food dispersion using Glander's (1975) descriptive data and his (Glander 1978) measurements of chemical composition and quality of the top 25 might be sufficiently resolved for bootstrapping analyses combined with global (population) parameters (life history and climate: Jones 1997a) describing events at the same site. A more informative approach, agent-based modeling, might utilize the aforementioned quanta with empirical data collected from foraging subgroups of a second mantled howler group (Jones 1980) located not far from the home range of Glander's subjects. These, and other, techniques (scaling, principal components, biophysical models) enhanced by precisely delineated verbal models have the potential to quantitatively describe abiotic and biotic factors driving behavioral sequences of mammals interacting with heterogeneously distributed limiting resources (food, water, breeding sites; Fig. 3.1), providing spatial and, particularly, temporal information applicable to other herbivorous mammals and to programs designed to mitigate deleterious effects to animal populations from environmental fluctuations, uncertainty, and risk. Comparable scientific data on microhabitats and foraging subgroups are available for other mammals (Bodmer 1990; Conradt 1998). Social groups and subgroups are breeding-grounds for exploitation; however, empirical reports suggest that benefits from membership may derive, on average, from flexible and energy-efficient access to limiting resources (Belovsky 1984).

Chapter 4
Robustness and Polyphenisms in Mammals: "Core Processes," "Repatterning," "Constrained Variation," and "Regulatory Logic"

"Given that phenotypic plasticity is obtained gratis, as a byproduct of the physics and chemistry of development, evolution of this plasticity can occur in two directions: One results in stabilization of the phenotype, effectively eliminating the plasticity [robustness], whereas the other results in the exploitation of the plasticity [polyphenisms]."

Nijhout (2003a)

"The limits to metabolic plasticity could be set by the production of reactive oxygen...leading to cellular damage, limits to substrate availability in mitochondria, and a disproportionally large increase in proton leak over ATP production."

Seebacher et al. (2010)

"Genetic variation is pervasive, but its expression as phenotypic variation is context dependent and heavily buffered."

Rutherford (2000)

Keywords Flück's Model • Polyphenisms • Atelidae • Carnivores • Primates • "Rapid" evolution

Among the unresolved questions in behavioral ecology and evolutionary biology are the origins and maintenance of biological diversity with evolutionary potential, topics that have been investigated theoretically and empirically at least since Woltereck's (1909 in Gotthard and Nylin 1995) research on three ecotypes of the freshwater crustacean, daphnia. Relative to its mean thermal niche, organisms may experience perturbations in one or more of three ways (1) increased or (2) decreased variability (frequency, rate, duration, intensity, type, quality, etc.) of endogenous or exogenous events (ambient temperatures, habitat fragmentation, drought), and/or (3) exposed to events novel to the repertoire of responses available to an organism (exposure to a new parasite or competition with an invasive species). Confronted with (endogenous or exogenous) perturbations, the organism must "decide"

C.B. Jones, *Robustness, Plasticity, and Evolvability in Mammals: A Thermal Niche Approach*, SpringerBriefs in Evolutionary Biology, DOI 10.1007/978-1-4614-3885-4_4, © Clara B. Jones 2012

("Hebbian") whether a change is "permanent" ("novel": Slobodkin and Rapoport 1974), an assessment process that may, in itself, be stressful, risky, and unpredictable. Slobodkin and Rapoport (1974) suggested that it is "parsimonious" to consider a perturbation "permanent" if it is one effecting stress at all levels of biological, molecular to phenotype (shock, "knock-down," Box 2.1).

Numerous authors (Rutherford 2000; Nijhout 2003a; West-Eberhard 2003; Kaneko 2009) discuss robustness and plasticity in the context of developmental events directed by physiological mediators (molecular, cellular, genetic, epigenetic, learned [acquired] responses [habits]). Molecular and genetic processes are usually treated as constraints on the developmental trajectory, buffering structures, and functions of development from the potentially disruptive effects of systemic insults. Abiotic environmental heterogeneity, in particular, changing climate and associated properties (*temperature*, humidity), is discussed ubiquitously as the primary source of variation inducing change in an organism's phenotype exposed to the external environment, potentially resulting in modified developmental processes ("developmental plasticity") and "genetic assimilation."

Considering each level of biological organization, clear distinctions among robustness and the two classes of plasticity, polymorphism and polyphenism, is not a straightforward challenge. Compared to plastic phenomena, robust states resistant to system perturbations are pervasive and relatively well-studied from molecular to ethological levels of organization. For example, many proteins are conserved across taxa. Molecular and cellular mechanisms preventing their spatiotemporal expression, dislocation, or displacement (structural or functional re-organizations of chemical pathways or neural networks) have evolved. In eukaryotes, two "fundamental strategies" repair errors in gene transcription ("transcriptional nonsense": Cusack et al. 2011; Wilkie 2011), splicing, translation, and post-translational modifications: "machinery" enhancing gene-expression efficiency (managing error rates) and "machinery" increasing the accuracy of molecular and cellular functions (minimizing consequences of potentially lethal "stress-loads").

Discussing amino acids responsible for premature termination of transcription events, Cusack et al. (2011) classify codons as fragile (constructing fragile amino acids), robust (constructing robust amino acids), or facultative (constructing either fragile or robust amino acids). Because rules governing the construction of amino acids function differentially, they are particularly subject to lethal errors, effects that can be minimized by the presence of robust amino acids in a protein transcription sequence. Thus, these error-correcting processes depend upon the tolerance of proteins for amino acid substitutions. Addressing problems of inference associated with correlations reported by Cusack et al. (2011) and issues concerning determinants of protein tolerances (thresholds, ranges of response, hierarchies, and probabilities of response) are beyond the scope of the present review. Nonetheless, the phenomena detailed by Cusack and his colleagues highlight the capacity for robust elements to generate plastic in association with other robust units of information at other levels of biological organization (physiology).

Robust states resistant to system perturbations and plastic states varying in response to endogenous or exogenous fluctuations may be induced by the same protein responding differentially to other chemical processes. For example, Dantzer and Swanson (2011) showed that upregulation (increased production) and down-regulation (decreased production) of insulin-like growth factor-1 (IGF-1), a hormone implicated in metabolic processes, covary with life-history features, specifically, growth, size, survival, and reproduction. These individual characteristics are molecular and cellular effects mediated by condition-dependent physiological and developmental mechanisms. Dantzer and Swanson (2011) advanced IGF-1 as a candidate mechanism underlying "the integration of life history and other phenotypic traits," exemplifying one thesis of the present review, that phenotypes at all levels of biological organization obtain from conserved "switches" and "toolkits" (Bell and Robinson 2011), including suites of connectomes, neural networks, circuits, modules, or programs (Sporns 2011).

Sharma et al. (1992) discussed "chemotherapeutic" properties of mammalian cells, probably originating as proteins protecting organisms from dietary toxins. Upregulation and downregulation stabilize mammalian cells, inducing "overexpression and amplification" of target gene products (peptides or proteins) or effects in the opposite direction, respectively, resulting in resistance to potentially lethal conditions. Responding differentially to stimulus inputs, regulatory mechanisms induced robust or plastic effects by mediating molecular and cellular processes. Structurally and functionally, this general ("classic") program, whereby expression of robust or plastic responses depends upon conditional signals communicating overriding system requirements, represents flexible biophysical dynamics essential to the continuing diversification as well as global ecological persistence and successes of terrestrial mammals.

Pasque et al. (2011) review in vitro studies of mechanisms responsible for increasing the efficiency of cellular reprogramming in mammals, emphasizing, in particular, reports of factors restricting "the natural reprogramming mechanisms of eggs and oocytes," such as the potential for epigenetic events to "resist" reprogramming of mammalian nuclei (Fig. 2.4). Studying "cross-resistance" to physiological stressors by *Drosophila* in the laboratory, Bubliy et al. (2011) concluded that the "evolution of shared protective systems associated with plastic responses may be constrained." These results suggest that biological systems are differentially sensitive to variations in frequency, rate, duration, intensity, and type of perturbation, demonstrating a need for additional research on mechanisms of reprogramming, including the extent to which they are recurrent and conserved within and among taxa. For example, constraint at the protein level does not necessarily predict responses to stressors at other levels of organization (physiological, ethological). Studying thermal tolerances in wild rat (*Rattus fuscipes*) populations separated geographically, Glanville et al. (2011) demonstrated differential physiological acclimatization to temperature between populations, but it was not possible to determine whether physiological plasticity translated to population resilience.

4.1   Laboratory Studies of Mammals Can Contribute to Understanding of Robustness and Plasticity: Flück's Research Program as a Model

Following Eisenberg (1981), wherever they occur, on land and in waterways, nonvolant mammals are the swiftest and most efficient vertebrate predators and prey, dependent upon neuromuscular, including phalangeal, modifications facilitating speed and deterrence of predators (hooves), navigation (flippers), capture and manipulation of prey (claws, opposable thumbs), food-processing (cooking), elaborate habitat modification (dams, burrows, extirpation of con- and contraspecifics, deforestation), construction (nests, houses), and manufacture of artifacts (tools). Within and between mammal orders, these characteristics are associated with flexible, often opportunistic, responses to unpredictable, fleeting, dangerous, risky, or other stressful environmental conditions potentially influencing survival and reproductive success.

Effects of Gene×Environment interactions on phenotypes are currently addressed in the laboratory with animals exhibiting highly conserved genomes (*Nematostella*: A. Reitzel, personal communication) and with standard mammal models (*Mus*: Keane et al. 2011), and Flück (2006) investigated "malleability" (neuromuscular plasticity) of mammalian striated muscle as a model of mechanisms underlying activity-dependent plasticity. Flück (2006) revealed covariations among structural (biophysical, "mechanics" of muscular activity), functional (metabolism, physiology), and gene transcription states capable of reprogramming ("remodeling," reorganizing, repatterning, plasticity) muscle tissues in response to phenotypic stimuli. Flück's (2006) ongoing research goal is to specify a molecular switch mechanism unifying processes of muscular remodeling (Box 4.1), a research project related to the investigation of how plasticity relates to the formation and maintenance of functional circuits. Hypothesizing a control mechanism driven by up- and downregulation (molecular feedback regulation), this researcher conceptualized his research questions in terms of theoretical biophysics, in particular, the relationship between variations in stress (exercise, load) and variations in strain (resistance to fatigue, elasticity) measured as "differentiation of muscle fibers" (mass, contractility, oxidative capacity). In brief, Flück (2006) is investigating the depletion of elasticity as stress increases toward its limit ("point of fracture": Roylance 2000, 2001).

Flück (2006) reveals how recurrent ("redundant": Kirschner and Gerhart 1998) metabolic responses to stress and strain are "co-regulated" by molecular and cellular processes, in particular, covariations between gene expression and metabolic (oxidative) functions. Flück (2006) concluded that mechanisms of skeletal muscular plasticity represent a unifying model for the in vivo study of muscular plasticity in mammals and other organisms. Several lines of evidence support Flück's (2006) optimism that conserved mammalian features comprising "toolkits" for expression and diversification of genetic switch mechanisms are identifiable. Supporting Flück's schema, Aziz et al. (2010) describe a "tissue-specific" master regulator of skeletal muscle gene expression whereby three conserved proteins work in concert

Box 4.1

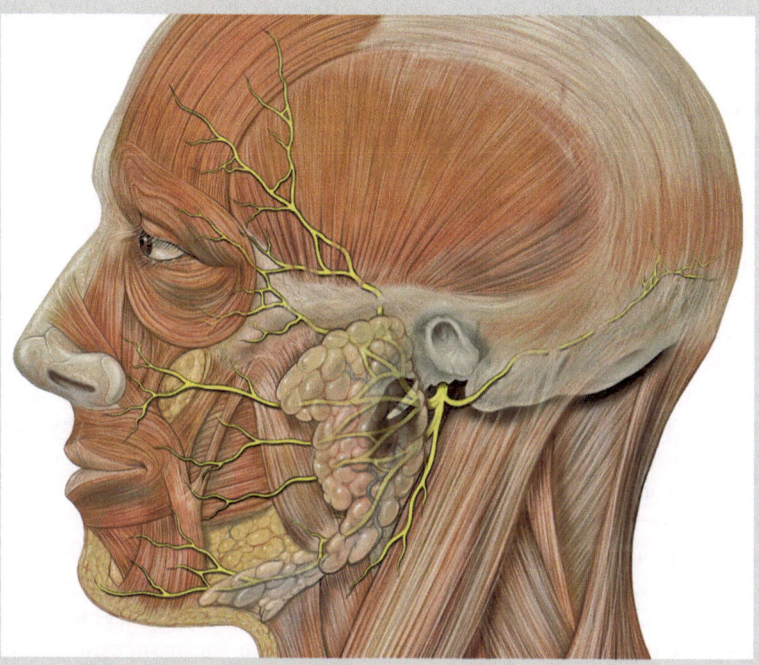

This illustration schematizes neuromuscular elements of the human face (Jones 2008) capable of discrete and continuous expression (Eibl-Eibesfeldt 1989), alone or in combination with other *functionally compatible* neuromuscular circuits throughout the body. For all mammals, "ongoing activity fluctuations" comprise "hypervariable components" ("hyperflexible," "exploratory systems," "unstable," "topological connectivity") exhibiting characteristic neural and chemical patterns (frequency, rate, duration, intensity) with communication (emotion) or other (structural support) roles. "Ongoing activity fluctuations" may be reproductively salient as "tinkering" mechanisms with the potential to generate biological novelty, diversity, and complexity, analogous, in some ways, to the operations of "transposable elements" (Jordan 2006; Jones 2008). As well, "ongoing activity fluctuations," via recurrent neural enhancement, may effect "long-term potentiation," reprogramming, and consolidation of biological innovations, including memories.

In humans and other social mammals, social (group foraging) and nonsocial (mating) contexts are likely to have favored the evolution of novel neuromuscular elements and neuromuscular networks (circuits) for efficient communication and intraindividual signatures ("personality," "syndromes"), and the neuromuscular

(continued)

Box 4.1 (continued)

inputs and/or outputs associated with these different contexts may conflict (exploitation, "response interference," "response competition"). Despite costs attendant to variability of biological· systems, combinations and recombinations of neuromuscular elements may enhance individuals' inclusive fitness (Gibb et al. 2011; Proppe et al. 2011; cf. Jones 2008, pp 48–49) and, like phenotypic variability generally, rates of neuromuscular synapse evolution may be differentially subject to and their phenotypic expressions differentially exposed to (West-Eberhard 2005; Piersma and Drent 2003; Jones 2006; Gotthard and Nylin 1995) selection (but see Sniegowski and Murphy 2006).

Stereotyped displays, classical ethological constructs, may be examples of neuromuscular elements originally favored by "behavioral accommodation" that, through recurrent exposure to exogenous stimulation, may favor "developmental bias" and "genetic accommodation" (Jones 2006, 2008). Several problems pertain to these formulations advanced, in particular, by West-Eberhard (2005), however, such as the likely requirement that reinforcement or reward is necessary to effect long-term potentiation (canalization, "behavioral accommodation," "genetic fixation," "genetic assimilation": Chap. 5) and genetic fixation ("genetic assimilation," "genetic accommodation") and that "schedules of reinforcement" (differential temporal features of reward occurrence: Chap. 5) may influence differential costs, benefits, and likelihoods of response of robust and plastic traits. Furthermore, under stressful regimes, alternative (synaptic) "trajectories," including ones already consolidated, may be more (reproductively) beneficial to individuals, singly or recombined, than novel trajectories. Like other propositions pertaining to potential benefits to individuals from variable phenotypes (Pigliucci 2008), transitions from an existing state to another in response to stress, at whatever level(s) of biological organization, rely on "rapid [rates of] evolution," a topic needing comparative re-evaluation (Evans et al. 2012). In a theoretical treatment, Scott-Phillips et al. (2012) explored a number of questions salient to the evolution of displays and attendant topics, particularly, conditions in which initial states (states prior to perturbing ones) are an ESS.

as a "mosaic," differentially changing, regulator of gene expression (MyoD), the signaling pathway from stimuli to fine-tuned muscular activity or its repression (negative and positive feedback regulation: Chap. 1).

Highly conserved regulatory mechanisms may be identified across animals. Mosaic transcriptomes characterized as a hypothetical "toolkit" are also the focus of research on eusocial social insects in Gene Robinson's lab (Chandrasekaran et al. 2011; Woodard et al. 2011). In a series of elegant empirical studies, these investigators have identified sets of conserved, rapidly evolving genes associated with signal transduction, physiology (gland development), and carbohydrate metabolism in three bee lineages (*Apis mellifera*). Compared to available evidence for bees and some other invertebrates, descriptions of regulatory signal pathways have not been described in detail for any vertebrate function. Progress toward this end is evident from recent reports detailing regulation of gene to phenotype expression in mammals (activity levels: Dayan and Cohen 2011; Korb and Finkbeiner 2011; organ function: Brawand et al. 2011; hair follicles: Wu and Zhang 2011; genome differentiation: Keane et al. 2011). Francis et al. (2003), studying differential expression of activity levels (variations in motor patterns, behavior) in mice, attributed their results to epigenetic processes rather than a conserved "mosaic" of regulatory proteins, and Brawand et al. (2011) as well as Wu and Zhang (2011) concluded that their findings could be explained by purifying selection, considered by many biologists to be the primary mechanism of "rapid evolution" in heterogeneous regimes. Other studies have shown that developmental events and cellular signaling networks strongly influence protein regulation, indicating, further, that the complexity of translation and transcription, their initiators and effects, remains to be dissected and diagnosed.

The latter studies suggest that gene expression may be mediated by more than one stress-responsive mechanism, neither compromising negative and positive feedback regulation nor obviating Flück's (2006) conceptual framework that biological systems may be constrained by perturbations induced by stressors (fluctuations in climate, nutrition) beyond the system's thresholds of response ("point of fracture"). To the contrary, alternate mechanisms provide additional sources of information for the investigation of endogenous and exogenous causes and consequences of the functional and evolutionary significance of regulatory diversity or of observed, recurrent deviations from conserved regulatory mechanisms, suggesting that costs of conservative features and their associated proteins (costs to efficiency, disease protection, or resilience) may enhance the evolutionary potential of other functional apparati, in particular, epigenetics (Fig. 2.4). Indeed, some questions are more readily addressed with the aforementioned evidence, including details of "crosstalk" between translation and transcription, as well as "rapid evolution" initiated by stress (in Flück's model) on muscular tissues. Section 4.2 discusses variability of mammalian motor patterns targeting the neck, many of these presenting as ritualized responses probably derived from an ancestral yawn reflex. Consistent with the ethological literature (Ewer 1968; Eisenberg 1981; Maynard Smith and Harper 2003), the following section's treatment strongly suggests that stereotyped, ritualized, and, often, species-typical signals derive from one or another suite of biophysical responses (reflex, spinal, autonomic) comprising regulatory mammalian circuits ("built-in units" Ewer 1968; Dumont et al. 2011) controlling "Hebbian synapse" networks via a brain's motor pathways. Flück's (2006) research program, and similarly others, could be expanded to test hypotheses concerning the coupling of robust (molecular) and plastic (behavioral) characteristics, including differential thresholds and responses to composite stressors varying in frequency, rate, duration, intensity, quality, and type.

Subgroups of structured animal populations are generally viewed as patches in temporally and spatially heterogeneous regimes (Emlen 1973). In many mammalian species, recruitment of offspring produced within a group is associated with dispersal and flexible reproductive rates in a regulatory feedback process inducing variations in group size and composition (adult and operational sex ratios: Emlen and Oring 1977) within and between years (Hanski and Gilpin 1997). Polygynandrous mantled (*Alouatta palliata*) and polygynous red (*A. seniculus*) howler monkeys, like most mammals (Durban and Pitman 2011), exhibit effective modes of short-, mid-, and/or long-range locomotion and movement, including dispersal and colonization, activity patterns that may vary with variations in environmental factors (habitat disturbance, temperature regulation, food availability and patch quality, sex ratio, flexible, reversible, and totipotent phenotypes as well as conserved RNA thermosensors: Meyer et al. 2011), including control of events prior to transcription, skin and hypothalamus (a structure located in the diencephalon (forebrain) controlling the mammalian pituitary ("Master Gland") and other functions critical to survival, thermoregulation, reproduction, and extraction of energy from food for differential biophysical allocation).

Studying small endotherms (*Rattus fuscipes*) during winter, Glanville and Seebacher (2010) measured activity levels and increased energy and food requirements, finding that fluctuating metabolic, particularly, mitochondrial, processes "sustained" biophysical responses to cold by "buffering" the rodents' phenotypes via feedback loops responsive to increased energy and food requirements when body temperatures fell below TNZ. Significantly, these authors demonstrated that, on average, lower diurnal activity (energy expenditure) was associated with "muted" mitochondrial responses to ambient temperatures, mitigated by increased activity levels (locomotion) mediating metabolic gene functions. Because mammalian females generally manifest lower mean basal metabolic rates compared to males in comparable conditions, the less responsive (and, more canalized?) sex experiences a relative disadvantage from cold stress and consequent energy demands of feedback regulation.

Investigating correspondences between diet and glucose metabolism in nectar-feeding bats (*Glossophaga soricina*), Kelm et al. (2011) reported that increased activity was more likely to benefit males with a higher initial activity rate compared to females in the same conditions by its enhancing effects on a male's ability to survive on a low-quality, high-sugar diet. The foregoing studies strongly suggest that female thermal tolerances are, ceteris paribus, more constrained than males', a metabolic profile that may predict significant sexual dimorphism in temperature-sensitive "switch" mechanisms benefiting male life histories (time-minimizing) compared to females (energy-maximizers) in the same population. Furthermore, if male mammals are characterized by flexible metabolic, particularly, thermosensory-mitochondrial, molecular, and cellular, regulation, male, and female mammals are expected to adopt segregated (binary?) life-history tactics and strategies, differential patterns of mean reaction norms required and permitted by males' greater responsiveness to and tolerance of low-quality food items. Because of higher thermal tolerances, male mammals, ceteris paribus, are expected to be regimented by fewer ecological constraints compared to females, a perspective consistent with pervasive sexual segregation among species of this Class. These propositions do not disallow flexible tactics and strategies among some females in some mammalian taxa in some micro- or macro-thermal regimes, but these propositions predict, instead, that flexible repertoires (whatever the level of biological organization) are likely to be more expensive energetically for females than for males in similar conditions.

Ceteris paribus, for mammalian males, higher variance in reproductive success from conversion of energy into offspring and, indirectly, grand-offspring, compared to females, is probably a greater function of differential metabolic profiles as suggested here than of anisogamy, per se, at least for taxa (or individuals, including phenogroups [units of organisms exhibiting overlapping reaction norms] and, possibly, guilds [interspecific feeding associations]) exhibiting relatively generalized and/ or flexible phenotypes. Ubiquitously, within and between mammalian taxa, males display relatively higher BMR and activity levels (male–male competition, including energetic demands of potentially "damaging aggression," mating effort, exaggerated signals and displays) compared to females in the same population, recommending the present, testable, topics as important research agendas with implications for causes and consequences of proximate and ultimate, sexually dimorphic, thermal feedback regulation in vertebrates (Boxes 2.1 and 3.1; Figs. 2.2, 3.1, and 4.2). McNab (1980, 1986) provides a very good entry into topics covered in this section.

4.2 Mammalian Motor Patterns Targeting the Neck: The "Regulatory Logic" of Molecules and Cells Coupling Robustness and Plasticity

Within and between organisms and populations, communication between at least one sender and at least one receiver functions as the sensory "glue" uniting all levels of biological organization, from molecules to whole organisms (Heidel et al. 2011). Signal design (structure) is generally conserved at the species level (Proppe et al. 2011), reflecting historical effects of environmental fluctuations (T). Species-typical signals may exhibit noteworthy condition-dependence (Proppe et al. 2011), exhibiting phenotypic variation controlled by genetic "switch" mechanisms (Handford and Nottebohm 1976). Among vertebrates as well as invertebrates, multimodal signals may be employed (Jones and van Cantfort 2007), and, for some mammals, functionally referential signaling about external events (Seyfarth et al. 1980) and context-dependent "meaning" (Quattara et al. 2009) have been demonstrated. Bradbury and Vehrencamp (1998) provided an exhaustive review of the physical properties, structures, functions, ecology, and evolution of animal communication mechanisms, including their energetic advantages and disadvantages, addressing differential costs and benefits of communication in terms of energy-expenditure per unit time (Parker 1974; Maynard Smith and Harper 2003). Other researchers have shown that communication systems may "buffer" within- and between-population interactions so that conflicts over limiting resources (food, mates, space) are rarely resolved by "damaging" aggression (Wilson 1975; Jones 1983). Studying black and gold howler monkeys (*Alouatta caraya*), Jones (1983) arranged phenotypic characters along a continuum from low to high likelihoods of energy-expenditure, permitting quantitative analysis in relation to individual, group, spatiotemporal, ecological, and other factors. Despite a very large literature on the structures, functions, and contexts of signaling within and between populations and species, Gotthard and Nylin (1995) advanced the *caveat*: "There are severe problems in deciding when a trait conforms to design specifications."

Maynard Smith and Harper (2003, Enquist et al. 2010) discussed the modification of generalized traits for communication function, processes generally associated with coevolved signals displaying "true communication" between signaler and receiver (Bradbury and Vehrencamp 1998). Tinbergen (1952) described the ubiquitous "ritualization" process whereby features of the conserved ("primitive") species-typical repertoire become "emancipated" or disassociated over evolutionary time from their previous function(s) leading to involuntary, exaggerated, simplified motor patterns known as "stereotyped" signals often termed postures, displays, or "fixed-action patterns." It is through the process of emancipation that stereotyped signals are said to be "derived" from earlier, conserved morphological elements. Stereotyped motor patterns are characterized by "typical intensity," relatively invariant and relatively immobile morphological features characteristically similar across individuals of the same sex and species. Finally, ritualized behaviors may present as "intention movements" whereby the stereotyped action pattern is identifiable but appears truncated and incomplete in form.

Employing or not employing an exaggerated, stereotyped signal is not, by definition, an intentional or conscious and aware "decision" since the evolutionary process leading to ritualization of phenotypic characters yields responses under autonomic control and resistant to endogenous or exogenous modification (Morris 1957; Lorenz 1981; Eibl-Eibesfeldt 1989). As a *caveat*, although there are certain mechanistic similarities, stereotyped motor patterns should not be confused with reflexes which are not ritualized, derived signals but action patterns induced at the level of the spinal cord controlled by the autonomic nervous system. In ethological terms, displays have evolved in response to particular stimuli termed "releasers," while reflexes have no particular communicative functions and are not stimulus-specific. Thus, rapid jerking of a body part from an aversive stimulus may be exhibited in response to a wide range of different stimuli such as heat, electric shock, or disgust. Ritualization leading to stereotyped phenotypic elements yields discrete (digital), discontinuous signals rather than graded (analog) signals transmitted as a continuous sequence of information-transmission (Eibl-Eibesfeldt 1989; Maynard Smith and Harper 2003; Enquist et al. 2010).

4.2.1 Mammalian Motor Patterns Targeting the Neck

The "neck-bite" is common to mammalian hunt and kill tactics and strategies, and predators and their prey are presumed to have coevolved, with predator hunting strategies a function of prey (plant or animal) defense responses (Sih 1985). Selection will be particularly sensitive to the relative sizes of predator and prey, favoring flexible responses, since hunting and defense are often opportunistic-, species-, and state-dependent (McCoy et al. 2011). Documenting the history, likely origin(s), forms, distributions, and functions of the mammalian neck-bite permits historical, comparative dissection and diagnosis of an ancestral motor pattern exhibited in monotremes and therians. The following background assembles information on motor patterns targeting the neck in mammals, forming a body of evidence from the scientific literature (Estes 1992; Eisenberg 1981; Swedell and Schrier 2009).

4.2.2 Carnivores

The mammalian literature on carnivore motor patterns targeting the neck primarily discusses these responses as they function during the kill. These morphological traits have been studied, particularly, in canids (wolf, fox, coyote) and felids (lion, tiger, leopard). Fossil carnivores are dated to the Eocene, ~55 mbp, and their common ancestor can be traced to the early Paleocene, ~60 mbp. Meachen-Samuels (2012) studied morphological convergence in extinct felids, finding a "positive functional relationship between saber elongation and forelimb robustness." The previous author concluded that the latter trait evolved in response to increased specialization and elongation of sabers, resulting in fragile (inefficient) structures when employed to subdue prey. Meachen-Samuels (2012) did not consider the possibility that strong, directional selection by sex may have favored the exaggerated, cumbersome sabers, a possible factor in declining growth rates of sabertooth populations. Since information is available about ecological and other environmental events during sabertooth tenure, it should be possible to estimate intrapopulation competition for limiting resources, particularly, mates, accounting for intense sexual selection on these displays in addition to factors explaining when and under what conditions the costs of unwieldy sabers outweighed their benefits.

The felid-canid split was completed in the Oligocene–Miocene, ~20 Mya, with felids splitting from the common ancestor earlier than canids. On the whole, these carnivore families evolved different patterns of hunting and killing though both groups primarily target the neck. Canids are characterized by group hunting, usually killing with a "death-shake" rather than the suffocating "neck-bite" employed by felids. Extant felids are solitary hunters generally larger than canids, character states likely to have favored a selfish and efficient, though individually risky, kill tactic. Compared to group hunting, solitary hunting is probably more sensitive to relative size differences between predator and prey. Solitary hunting and the neck-bite are probably the derived states compared to group-hunting and the death-shake since the neck-bite is more energetically efficient. These inferences are supported by results from canid-felid morphological comparisons whereby canid morphology is the more robust and generalized, and felid morphology the more derived and specialized (Meachen-Samuels and Van Valkenburgh 2009), relationships that these authors attributed to different "prey-size preferences" (cf. results for cheetah, p 741). This scenario is consistent also with reports that the felid line is older than that of canids, since felids would have had a temporally wider window for evolutionary changes to occur.

While characteristic hunt and kill tactics and strategies have been documented for carnivores, there are exceptions to observed patterns awaiting detailed analysis. For example, among felids, jaguars and cougars, Neotropical taxa not closely related to each other, preferentially kill by using their teeth and jaws to crush the temporal lobes of prey, a method dispensing of prey virtually immediately compared to canid and other felid kill techniques, possibly neutralizing advantages gained from fore-limb robustness and contrasting with patterns reported in the 2009 article. Similar to the aforementioned comparisons between canid and felid kill-bites, jaguar and cougar skull-crush motor patterns are relatively efficient. The mammalian litera-ture generally holds that skull crushing is an adaptation favored during the Pleistocene's great mammal extinctions when many available prey items were armored reptiles, a hypothesis seemingly supported by the observation that, while not a particularly large cat, cougars feed upon large, dangerous prey such as moose, elk, and bighorn sheep.

With the exception of humans and bonobos, in addition to gorillas in captivity, one or another variation of the dorsal-ventral sexual posture is ubiquitous among mammals. Male canids and felids employ a stereotyped neck-bite during copulation, interpreted as a character trait derived from ancestral kill states. Neck-bite variations displayed during mating function as intention movements resulting in little or no harm to a female *in copula* since the force of a bite to the point of serious or lethal damage is inhibited. During sexual congress, some male mammals use neck-bite motor patterns to maintain balance during mount and intromission phases and, by inference from published literature on these topics, stereotyped neck-bites may function multi-modally, for example as sexual and aggressive communication to females and other males.

Other ritualized neck-bite responses apparently derived from an ancestral toolbox occur in association with maternal "mouth transport" and play. Females in many orders of mammals move their young from one location to another using a stereotyped method of transport employing mouth and teeth with some force, sufficiently inhib-ited and truncated to prevent harming young, though elements of the motor pattern may communicate the female's "motivation" (aggression, "contact comfort," punish-ment). Similarly, the stereotyped play-bite, characteristic of most social mammals, is proposed as a derivation from an ancestral anatomical and neuromuscular toolkit. A number of functions have been attributed to play, but most reports emphasize its role in intraspecific social bonding, as an interaction permitting assessment of relative strength, or as practice for responses common to adult behavioral repertoires.

4.2.3 Primates

With the exception of three apes (bonobos: *Pan paniscus*; chimpanzees: *P. troglodytes*; humans: *Homo sapiens*), coordinated hunting and killing of conspecific and contraspecific prey is rare or absent among primates. The three hominid taxa are pri-marily terrestrial, heavily dependent upon clumped plant resources, particularly, fruit, and widely acknowledged for their expert tool-use. In the nonhuman apes and a few anthropoids (capuchins), tools may be employed to break shells of nuts or crustaceans, to extract termites or ants from their colonies, or to extract marrow from bones and brains from skulls. Where killing has been observed in monkeys (Rhesus macaques: *Macaca mulatta*; capuchins: *Cebus capucinus*) opportunistic capture and kill tactics and strategies were employed, and food items included both invertebrate and vertebrate taxa, including ants, spiders, birds, small mammals, as well as conspecific and contraspecific anthropoids. Bonobos, chimpanzees, and monkeys generally kill prey by dismemberment or disembowelment, using jaws and teeth, facilitated by hands, in both contexts. A few reports document the use of skull crushing in the manner of cougars and jaguars to effect infanticide, and a media release recently reported that female Sumatran orangutans (*Pongo abelii*) may immobilize slow loris (*Nycticebus coucang*) with a swipe of the hand before dispatching the prosimians with a skull crush. Anecdotal accounts suggest that chimpanzees may crush the skulls of their victims, including the occasional adult female by an adult male. It will be interesting to assess sexual dimorphism in hunting since, in some taxa, males are most likely to hunt (capuchins, humans), in others, females (lions, bonobos).

When primates kill, action patterns may appear intentional and graded (dispatching prey with tools) rather than discrete and stereotyped (death shake, neck-bite) response patterns suggestive of flexible genetic switch mechanisms, regulatory feedback processes that may not have been investigated for *alternative* kill methods (polymorphisms). Humans are noteworthy for employing all manner of opportunistic and "intentional" methods to kill conspecifics and contraspecifics, not only for food, but also legitimately (policing, self-defense) or by accident (vehicles, sports events), as well as for other reasons frequently classified as illegal and proscribed acts (murder, paraphilias). The relative robustness and plasticity of these phenotypes, including their phylogenetic origins, could, in theory, be diagnosed with some variant of the methods discussed in Sect. 4.1 and Synopsis. Cultural rituals incorporating or leading to death (cannibalism, human sacrifice) are not uncommon practices in human cultures, and my research indicates that the ethical status of these practices is an ongoing topic of debate among Western philosophers and clinicians.

In bisexual contexts among nonhuman primates, the stereotyped neck-bite or, similarly, a stereotyped shoulder-bite (Fig. 4.1) may or may not be displayed, even in species characterized by a detectable male dominance hierarchy. With the exception of bonobos and humans that frequently intromit in ventral–ventral orientation, primates generally copulate in the dorsal-ventral position like other mammals, with a male typically using his feet to restrain the female's ankles and to maintain balance during intromission, thrusting, and dismounting. Each of these sexual postures, particularly dorsal-ventral, exhibits stereotyped elements (typical intensity of torso, arms, facial gestures, and, in some species, vocalizations). Where neck- or, less frequently, shoulder-bite motor patterns occur in combination with sexual congress, usually post-orgasm, reports in the primate and other mammal literature rarely document damage to the female *in copula*, with the exception of lions: *Panthera leo*, domestic cats, Hamadryas baboons: *Papio hamadryas*, and humans. Hamadryas males aggressively "herd," coordinate, and control harem females; however, though the teeth of these males may pierce skin around females' necks, the motor pattern is ritualized because the act of biting to kill is inhibited and incomplete.

Fig. 4.1 The postcopulatory, stereotyped "shoulder-bite" (*Gorilla gorilla beringei*: lowland gorilla), a modified ("emancipated": Tinbergen 1952) "neck-bite." According to the researcher who documented the depicted motor pattern, this display always occurs during ejaculation. The female *in copula* may respond with screams or may chase the servicing male; at other times, however, the female remains silent and passive. Both the "neck-bite" and "shoulder-bite" occur in Hamadryas baboons (*Papio hamadryas*: Swedell and Schrier 2009), and anecdotal reports suggest that, on rare occasions, humans may exhibit similar behaviors. A review of the mammal literature shows that biting displays directed by males to females in mating contexts do not necessarily represent "coercion" or "force," male aggression toward females is not necessarily favored in all conditions, and that male aggression towards females is most likely to occur where females are or are suspected to have been promiscuous. These and related topics have not been comprehensively investigated in mammals. ©Rick Murphy

Where it has been reported to occur, the neck-bite is "gentle" in prosimians (greater bushbabies), most of whom are solitary and nocturnal. In general, males of conservative primate species are reported to be "patient" with females in estrus (lesser bushbabies), not attempting to mount and intromit until the female communicates receptivity behaviorally (approach, stereotyped rear-present posture, vocalizations). In humans, ritualized neck-bite motor patterns displayed in sexual contexts have not been reported in the scientific literature, though non-stereotyped "love bites" are common, particularly as components of flirting, play, sexual foreplay, paraphilias (sometimes leading to death), or other intimate contexts (mother–child interactions).

Play focused on the neck is virtually ubiquitous among primates and, similar to carnivores, may occur as stereotyped play-bites, particularly in nonhuman species. Primate play bouts are often reciprocal, with interactants soliciting play by vocalization, exposure of neck or other areas of the body, or touch (tickling). Based on my analysis of You Tube videos, play among anthropoids, especially cercopithecines, frequently includes apparently stereotyped play-bites. The stereotyped play-bite was not apparent in the hominid images investigated. Stereotyped components of play in nonhuman apes appeared to be solicitations to play, sometimes involving a bared-neck display. Otherwise, bouts of play did not appear biased in favor of the neck region. Grooming, the most common social behavior among nonhuman primates, indicated that monkeys groom most of the surface of a recipient's body, though hand postures appeared to include stereotyped elements. In my sample of anthropoids, grooming was not skewed in favor of the neck region. Except for humans, apes frequently targeted the neck during grooming. Solicitations to groom, but not grooming bouts, appeared to include stereotyped components.

Based on You Tube videos and photos posted online, human play directed at the neck and other bodily regions was primarily graded and intentional. Some action patterns exhibited during play bouts, flirting, or in association with lovemaking appeared to be "intention movements," exhibiting stereotyped elements (kissing, stroking, gentle scratching), all of these motor patterns share similarities with both play and grooming in other primates and in mammals generally. Primate phenotypes incorporate both ritualized and non-ritualized features, and this mosaic pattern of surface traits enhances flexible, condition-dependent communication. Stereotyped elements of intentional acts of play in the neck region are not uncommon in humans (snuggling, mild-nuzzling, "huddling," kissing, gentle biting, caretaker manipulation of offspring necks and faces using lips, teeth, or phalanges, sometimes to induce sucking at the breast).

4.2.4 Diversification of the "Neck-Bite" Motor Pattern in Mammals

Phylogenetic distribution of neck-bite motor patterns in carnivores and primates demonstrates functional and structural variations, supporting the hypothesis that an ancestral circuit regulated by protein switch mechanisms is amenable to rapid evolutionary pressures. For example, very similar neck-bite motor patterns are employed by carnivores among sexual contexts, and male monotremes occasionally employ the neck-bite during mating. On the other hand, modes of hunting and killing between canids and felids have diverged over evolutionary time. Other less common variations of lethal neck-bite motor patterns are found in carnivores such as the skull-crush and disembowelment, a behavioral sequence sometimes occurring in combination with the neck-bite (lions). Kill tactics and strategies exhibited by primates are roughly similar to those displayed by carnivores. Unlike canids and felids, no nonhuman primate species utilizes a ritualized neck-bite for lethal ends, although non-stereotyped, not necessarily lethal, neck-biting has been documented for extant humans in sexual contexts (paraphilias). These and other topics related to mammalian biophysical (motor patterns, behavior) circuits are amenable to investigation by research programs such as Flück's (2006; Sect. 4.1).

The previous review of ritualized mammalian neck-bite motor patterns demonstrates structural and functional divergence within and between orders over time. The scenarios described for carnivores and primates are consistent with explanations in the mainstream biological literature whereby ritualized signals originate with physiological and/or behavioral cues that are, subsequently, selected for signal functions. Among the categories of response giving rise to signals and displays (Maynard Smith and Harper 2003), yawning is a strong candidate as precursor to death-shake and neck-bite kill tactics, diversified over time in structure and function, in particular, disinhibition of lethal force and social operations, respectively. Yawning, and other motor patterns discussed, embodies additional features with potential signal function (teeth) that, in combination with other traits, might be employed for multimodal (tactile, visual) significance. Related to the incipient signaling functions of yawning, displaying teeth is generally a sign of aggressive motivation, while covered teeth represent nonaggressive motivation or subordinate status. It is interesting to note that the evolutionary pathways ("trajectories") from yawn (hypothesized) to kill-tactic, to mating element, to play reveal changes in motor patterns targeting the neck not only in force but also in likelihood of dental exposure, topics worthy of comparative phylogenetic and ecological investigations, in addition to physiological and energetic estimations. These and other observations in the present section are amenable to investigation by the model proposed by Flück (2006), and it will be of general interest for such a project to specify adaptive relationships between robust, stereotyped, signals and displays, on the one hand, and flexible, context-dependent expressions of behavior, on the other.

Maynard Smith and Harper (2003) discussed organisms' responses to novel stimuli, the latter's evolutionary redirection to neck-biting and subsequent ritualization. Typically, animals "generalize" from detectible features of a novel stimulus to stimuli comprising the existing repertoire, a well-studied psychological phenomenon. Assuming that novel and experienced stimuli are distinguishable, responses to the former may be subject to selection if existing responses are suboptimal and novel ones yield a higher optimum. This condition describes one confronting populations in nature whereby an environmental change, say, temperature, increases beyond some threshold of tolerance yet remains within a stimulus range detectable by sensors. Novel stimuli may also be arbitrary with respect to vital characteristics of a population's shifting optima. In these conditions, a population's responses to stress may be "dramatic" or "unpredictable" (Maynard Smith and Harper 2003), favoring "rapid evolution," assuming sufficient effective variation (Introduction, Sect. 4.1). Theoretically, any phenotypic variation, differentially and reliably activated in response to the same environmental properties, has potential to induce population divergence and genetic isolation (Chap. 5).

In the foregoing examples, females may be "attracted to" different male traits (different displays or other motor patterns associated with copulation) because of reliable differences in sensory modalities including differential force of male biting and other motor patterns displayed conditionally (intromission force, rates of thrusting during intercourse, "forced copulation," Sect. 4.4). The epigenetic effects (Fig. 2.4) of "female choice" deserve intense scrutiny following reports from insects that female selection on male traits induces responses compatible and complementary with the interests of females. In the present case, feedback regulation of a male's (sender, parasite, predator) neck-bite (coercion, force, and intensity) signals information to a female (receiver, host, prey), leading to decisions by the receiver's "Hebbian" synapses about how to respond based on probable consequences, determined, for example, by differential rewards (Chap. 5) consolidated in "neural memory" from past experiences (positive or negative reinforcement: "neural memories" generated by reduction in force, discomfort, pain, coercion). Continuing the hypothetical feedback trajectory, female responses (withdrawal, aggression, display) impose selection intensity on the sender favoring force-reduction, also favored by neural memories of reward (positive reinforcement: increased likelihood of female receptivity or, via male–male competition, access to females or resources preferred by females). In the extreme case, a male's neural memory (of reward insufficient to benefit male interests) may induce increased neck-bite force up to a point lethal to a female and reaching her "give-up-point" or threshold, transmitting information to a female's regulatory circuit to switch to a pure strategy of "kin selection," since an organism will "choose" death when the benefits incurred via indirect reproduction by decreasing competition with relatives outweigh benefits from continued (allocation to) direct reproduction (Ricklefs 1977).

Other stressful events, not limited to a signaler's bites, may induce heritable, hormonal responses in stress-sensitive regulatory processes (CRH in vertebrates), leading to strong selection on exaggerated traits, may cause differential recruitment of protein or neural pathways, activate coevolutionary (Box 3.1) and/or frequency-dependent processes, or may expose cryptic phenotypic variation. A range of possible mechanisms may drive "rapid evolution" (rapid fixation: Greenfield 2002; Schoener 2011; Slobodkin and Rapoport 1974; Jones 2006; Maynard Smith and Harper (2003); Tinbergen 1952; Hallgrímsson and Hall 2005), such as the biophysical features (anatomy, sensation, physiology) associated with male kill tactics previously discussed. Consistent with the present review's advancement of a thermal niche approach to evolvability (Chap. 3; Dell et al. 2011; Seebacher et al. 2010), Nijhout (2003a) proposed that ritualized communication and other processes (aging, evolution, cancer) may be regulated by an "insulin signaling network" integrating nutrition (energy inputs) and metabolism (mitochondrial functions).

Because of their energetic advantages over females, males are generally the primary beneficiaries of costly, ritualized indicators of resource-holding potential (strength, status, attractiveness, quality, health). However, mammalian displays are not typically assessed relative to differential reproductive benefits or benefits from enhanced access to resources (Chap. 5). A review of mammalian motor patterns incorporating the neck reveals a very broad array of stereotyped and non-stereotyped responses in addition to those discussed in this section ("play panting," the "neck present," "LEN facial expressions," the "social scratch," "abrasive" uses of the neck), and many of the derived, "emancipated" functions appear specialized for use by subordinate individuals, particularly, subordinates to dominants (Fig. 4.2), suggesting, as well, that neck exposure and manipulation of the neck (grooming, conflict, "contact comfort") may have derived from hygienic behaviors such as extirpation of parasites.

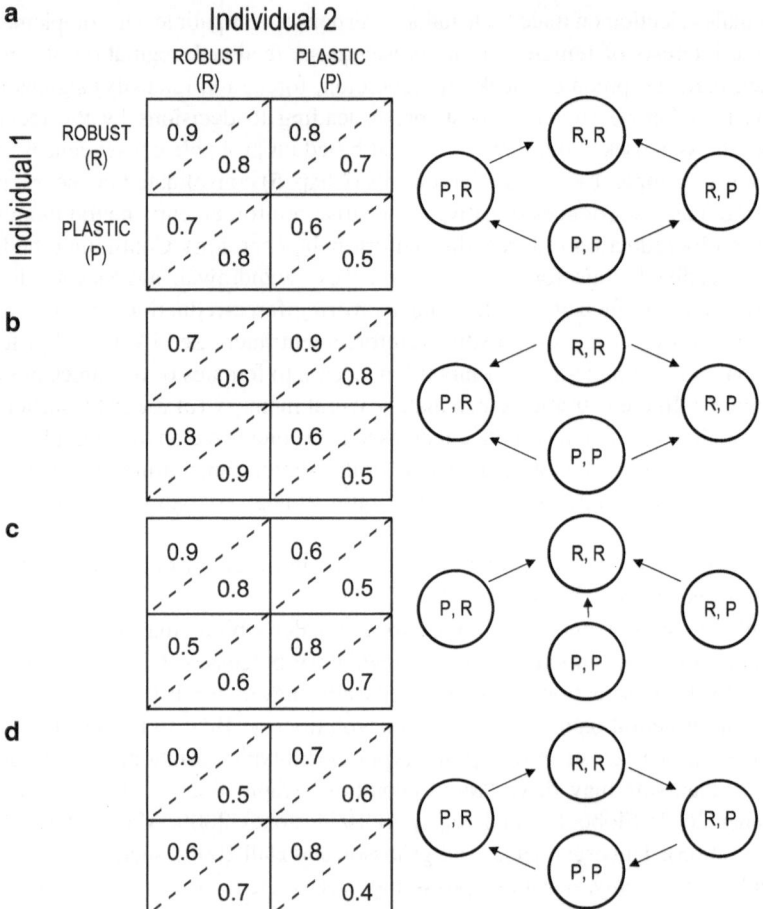

Fig. 4.2 A two-person game displayed as a typical payoff matrix in each of four (hypothetical) conditions. *Rows* and *columns* are controlled by individual 1 (dominant female) and individual 2 (subordinate female), respectively. Transition graphs are displayed to the right of each payoff matrix showing expected outcomes if individuals make decisions optimally. See text for further explanation and discussion. ©Clara B. Jones

4.3   Stereotyped Motor Patterns as Polymorphisms: An ESS Approach to Interactions Between Females of Different Dominance Rank

Gross (1996; Jones and Agoramoorthy 2003) argued that social interactions drive individual phenotypic variation within the sexes and that game theory and the evolutionarily stable strategy (ESS) concept may be used to analyze "how and why selection favors alternative phenotypes." Figure 4.2 displays a hypothetical two-individual (two female) game. Each female is presumed to act selfishly to optimize her inclusive fitness, and female 1 (relative fitness "payoff" above diagonal), with greater "resource-holding potential" (Parker 1974), is (in the initial state) dominant to female 2 (relative fitness "payoff" below diagonal). Following Gross (1996), social rank is assumed to be density- and/or frequency-dependent, possibly sensitive to interaction rates, and rank may vary spatiotemporally (a function of different reaction norms in different "patch" conditions [microclimates]). In general, dominants are assumed to accrue higher relative fitness for the period of time that the interactions persist. However, "role-reversal," whereby the subordinate wins a contest with the dominant and the dominant subsequently becomes subordinate or is ejected from the group, may occur where the subordinate has more to gain by winning than the dominant has to lose (Parker 1974; Jones and Agoramoorthy 2003; Horwich et al. 2001). Each of the following examples represents routes whereby stress (conflict, changed fitness optima, chemical suppression of ovulation) may induce variations in phenotype ("alternative phenotypes": coalitions and alliances, "self-restraint," dispersal, role-change or reversal), decisions made at "Hebbian" synapses induced by regulatory feedback mechanisms sensitive to the economics of individual response hierarchies, including the energetics of transition probabilities (Fig. 3.1).

As pointed out by Johnstone and Cant (1999), agonistic interactions between two individuals may have one of three outcomes (1) the loser may be killed (a likelihood minimized by ritualization, assessment mechanisms, and conventional displays), (2) the loser may be evicted from its group, or, (3) the loser becomes or remains subordinate. Figure 4.2 may also be interpreted as a game between female kin whose decisions will be sensitive to the intensity of competition in different environmental regimes (Perez-Tomé and Toro 1982). Figure 4.2 illustrates the straightforward example of contests between two females (access to an infant, to a mate, to food) using two different strategies (condition-dependent behavioral phenotypes: Gross 1996): a robust (R), stereotyped response (exposing ["neck-present" posture] or not exposing the neck's surface to the other individual) and a plastic (P) response (manual grooming in the neck region). If R, a female will initially benefit reproductively from a "neck-present" (higher payoff); if P, the opposite pattern obtains. The four matrices (Fig. 4.2a–d) display different (hypothetical) fitness values in four states (competitive regimes), and females' decisions (effected by "Hebbian" synapses) are presumed to reflect individual self-interest.

In each condition, four outcomes are possible: R, R, when both females decide to "neck-present," an outcome that should lower overt displays of agonism because both females benefit from using a stereotyped response (that, by definition, decreases likelihoods of aggression). In a second outcome, R, P, female 1 decides to use a stereotyped, restrained response ("neck-present"), and female 2, a plastic response (groom). Female 1's decision may obtain in order to minimize agonistic costs from direct competition with female 2 or to maximize benefits (increasing investments from the subordinate "helper"). It may, however, be in the interest of female 2 to challenge the social status of female 1 ("dominance testing," interference competition), a condition with the potential to generate alternative tactics and strategies by female 2 (coalitions or alliances with a male or with other females) even though "dominance testing" is expected to increase the likelihood of aggression. Decisions based on mutual assessment are likely to minimize direct or indirect agonism between females since, on average, the dominant, female 1, is expected to have a mean advantage in pairwise interactions.

For the third possible outcome, P, R, dominant female 1 benefits from grooming (plastic), while it is more beneficial for female 2 to "neck-present" (robust), a condition obtaining, for example, from condition-dependent (microclimate-dependent) variations in fitness optima decreasing asymmetry within the female dyad (higher predictability of limiting resources [food]). In such situations, female 2 has more to gain than female 1 has to lose (Parker 1974) with the consequence that female 2 may benefit from attempting to dominate female 1 with the more efficient display ("neck-present"). Where female 2 is in a position to challenge female 1's social rank, female 1 may benefit more from using a plastic response ("neck-present") providing a wider range of [alternative] responses. The fourth outcome, P, P, pertains when both females benefit from a decision to "neck-present," the more variable response. This choice may obtain where neither female stands to gain from "aggressive restraint" (the less variable, more robust, "neck-present"), for example, where environmental fluctuations severely stress an organism's access to limiting resources. Each of the preceding conditions is likely to depend upon several factors in addition to relative resource holding potential (mean relatedness with groups, ecological constraints, group size, number of adult females in a group, monopolizeability of limiting resources).

Figure 4.2 also displays a "transition graph" to the right of each matrix showing changes in decisions (at "Hebbian synapses": displayed within circles) based upon payoff values from one cell to another. In Fig. 4.2a, only strategy R, R is stable (no other strategy can invade it, as indicated by the convergence of arrows on R, R). In Fig. 4.2b, both P, R and R, P are stable (arrows converge on these two strategies). In Fig. 4.2c, only R, R is stable (convergence of arrows). Finally, Fig. 4.2d exhibits no stable equilibrium (arrows converge on no strategy). Following Pulliam and Caraco (1984; Gross 1996), transition graphs assume only that (1) Each individual has sampled the four possible states and has processed their relative benefits and (2) An individual switches her decision only if higher relative benefits can be obtained from doing so. Figure 4.2 thus displays "perfect [idealized rather than realized] information decision ['Hebbian'] rules."

The transition graphs can be described in greater detail. Figure 4.2a demonstrates equilibrium only when both females decide to exhibit a "neck-present" display. Thus, either female benefits from a switch to R from any other state. Once one female decides upon R, the other will do the same, since either stands to gain from such a choice (cf. fitness values in matrix). In the final result, once both females have decided on ("chooses") R, the condition is preserved as long as any further move results in lower relative inclusive fitness for either female. In Fig. 4.2b, the costs of R outweigh the benefits (R, R accrues a lower fitness value than R, P or P, R). Either female thus benefits from a decision to groom (P) when the other female is in state, R. Therefore, whenever both females are interacting, the only stable equilibrium occurs when one female chooses R, and the other, P (R, P or P, R).

Figure 4.2c displays a state comparable to Fig. 4.2a whereby the benefits of R outweigh the costs (the highest benefits to each female accrue when both exhibit a "neck-present" display [payoff matrix]). In this state, however, when both females choose P (P, P), R fails to pay more, as long as the other female retains her choice. However, both females benefit by a decision to display (R, R). Thus, following Pulliam and Caraco (1984), state R, R is a "cooperative" (Pareto) equilibrium. Figure 4.2d displays a case for which no stable equilibrium exists. The consequence of such an unstable equilibrium is a "Round-Robin" (Pulliam and Caraco 1984) whereby female 1 chases female 2 from situation to situation. As Pulliam and Caraco (1984) point out, such unstable interactions follow from optimal decision rules whenever one individual always benefits from one trade-off and the other always benefits from an alternate one (in the Fig. 4.2d matrix, no matter what decision one female makes, the other female can always do better with an alternate one). Conflicts of interindividual interest may be particularly intense under these conditions, possibly existing where "scramble" competition obtains (extremely unmonopolizable, unpredictable, fluctuating, risky, and/or inescapable stress; competitive conditions in relatively large groups: agricultural or industrial societies?). Such conditions may induce alternative responses such as the various forms of exploitation, division of labor, specialization, including castes, coercion, force, repression. Figure 4.2 provides proximate explanations for variations in the number of adult females in groups as well as related features of groups (adult sex ratio). Finally, states displayed in Fig. 4.2 (independent of the particular sexes or responses differentiating one individual from another) should be very sensitive to within-group coefficients of relatedness (r), especially for females. The propositions in this section are consistent with the verbal model advanced by Emlen and Oring (1977); however, these authors failed to factor r into their formulations.

4.4   Adaptive Polyphenism: "Spatiotemporal Compartmentalization" of Discontinuous Phenotypic Variation (Nijhout 2003a)

Conradt (1998) discussed ruminant (herbivore) species characterized by sexual segregation, proposing that this population structure arose from divergent patterns of food choice between males and females. Where potential mates are separated spatially, reproductive isolating mechanisms might arise by conventionally acknowledged mechanisms (geographical barriers, isolation by distance) if population subdivisions include viable numbers of reproductive males and females. Howler monkeys (*Alouatta* spp.), the most widely distributed Neotropical primate genus, has diversified into ~7 parapatric species (Groves 2001), from Mexico to Argentina. Contact zones between howler species occasionally occur at species boundaries, and viable hybrids apparently are produced. In Brazil, photographic and descriptive records strongly suggest that hybridization between formerly allopatric black and gold howler monkeys (*A. caraya*) and brown howlers (*A. guariba clamitans*) has occurred in forests degraded by anthropogenic effects (Bicca-Marques et al. 2008). Despite occasional sympatry within the genus, *Alouatta*, infrequent range overlap strongly suggests that species-isolating mechanisms have been effective.

Candidate isolating mechanisms can be hypothesized for both sexes. Adult females are characterized by genital hypertrophy, and qualitative observations suggest that vulva mass (and morphology?) varies significantly between subspecies (*A. p. palliata* and *A. p. mexicana*: Jones 1997d, 2002; Eberhard 1985) but not between species whose geographic ranges are widely separated (*A. p. palliata* and *A. caraya*: Jones 1997b). This pattern of results suggests three nonexclusive, testable hypotheses (1) Sensory bias facilitates character displacement (here, vulva mass) via "female choice," (2) Reinforcement (postmating, prezygotic effects: female mortality from parasites or aggression by conspecifics of either sex, hybrid inviability or infertility), (3) Spatiotemporal factors related to environmental differences (temperature, resource dispersion: Jones 1997a). Analyzing the first two hypotheses with quantitative techniques, Servedio (2001) concluded that postmating, prezygotic selection pressures account for most "preference divergence." Another suite of questions concerns the prominent, mottled-white scrota displayed by adult male howler monkeys (Jones 1999b), a possible case of "fluctuating asymmetry" (random deviations from symmetry of traits on opposite sides of the body: Møller and Pomiankowski 1994) of morphological structures caused by developmental instability.

Comparative, empirical research on water-strider clades (Gerridae) recently reported by Rowe and Arnqvist (2012; Eberhard 1985) demonstrated the relative importance of morphology and complexity over size for analyses of genital structures. Though the evolution of animal genitalia is "poorly understood" and complexity and divergence difficult to measure, two "trends" are evident from a review of the scientific literature (1) genital traits diverge more rapidly than nongenitalic ones, (2) structures are taxon-specific. Comparative analyses by the aforementioned investigators demonstrated that intromitting structures (penis of male mammal) exhibited greater complexity and divergence compared to nonintromitting genital structures (vulva mass of female mammals), correlated with postmating and premating "indices," respectively. Variability in genital complexity was an apparent consequence of sexual selection acting on each sex independently.

For reproductive females, male genital structures may function as indicators of male viability ("good genes") or attractiveness, sexually selected signals of "good genes" or "Fisherian" mechanisms, respectively (Greenfield 2002). Allen and Simmons (1996; Eberhard 1985), studying dung flies (Scathophagidae), found that fluctuating asymmetry in visual displays employed by males (for howler monkeys, variations in scrotal asymmetry and size: Jones 1999b) exhibited an "equivocal" association with male fitness compared to structures having "mechanical" significance (the penis for male mammals) and that both features were associated with "coercive mating" in their study species, providing another example of the utility of employing ethological conceptualizations of traits as discrete properties of phenotypes (Chap. 1, Sect. 4.2).

In primates and other mammals, sexual displays are generally multimodal (multichannel, multisensory, "complex," multinodal), combining, for example, audiovisual, olfactory-visual, tactile-auditory, signals. As reviewed by Jones and van Cantfort (2007) and Jones (2002), several scientists have proposed that multimodal signals increase the reliability of information sent to receivers and/or increase the ease of signal reception. It has also been suggested that multimodal signals and displays enhance "adaptive" information-processing, potentially enhancing the expression of novel responses, "exploratory systems," "hypervariability." "Complex" signals may also function as "multiple messages" or "redundant signals," supporting the view that neuromuscular elements (Box 4.1) are components of hyperflexible systems enhancing evolvability potential. These and related topics emphasized in the present review should apply to all "exaggerated" traits (ornaments, genitalia) among mammalian males, and recent evidence for females of the Class demonstrated that "sperm competition" and sexual conflict may lead to positive (coevolutionary: Box 3.1) selection on female features, as well ("reproductive proteins": Swanson et al. 2001).

Examining vertebrate research, Greenfield (2002) pointed out that biophysical correlates of motor patterns are likely to be heritable, with trait variability maintained in populations by Gene×Environment interactions (mean reaction norms). Since landscapes vary spatially (Greenfield 2002), it is possible that different reaction norms across a fluctuating thermal niche may facilitate population divergence and diversification of taxa via sexual selection (Maan and Seehausen 2011), mechanisms universal among mammals and other vertebrates (Jones and Agoramoorthy 2003). With probing verbal-graphical models, Bonduriansky (2011) advocated the view that sexual selection and sexual conflict facilitate ecological diversification by shifting fitness optima of populations, inducing individuals to "explore a wide phenotypic niche" ("tinkering," "exploratory processes," "hypervariability," evolvability: Kirschner and Gerhart 1998), a functional process that the author terms "ecological co-adaptation."

4.5 Benefits of Group Membership: "Information Centers," Flexible Access to Limiting Resources, and Phenotype Buffering

Lewontin (1957) posited that temporally fluctuating regimes favor homeostatic responses, and subgroup formation in response to unpredictable, ephemeral, or rare, limiting resources is one mechanism capable of regulating relative energetic costs and benefits. Subgroup formation may differentiate populations based on natural categories such as age, sex, pelage color or pattern, and dominance rank. The size and composition of groups may have important consequences for the survival and fecundity of organisms (Pulliam and Caraco 1984; Wilson 1975). A subgroup is defined here as a unit (>1) of a demographic group functioning in similar or different

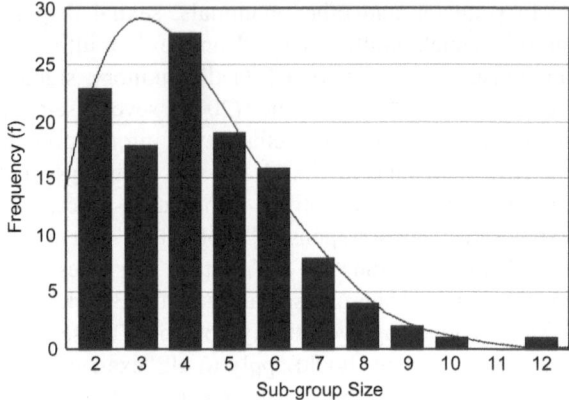

Fig. 4.3 Frequency and size of subgroups for one demographic group of Costa Rican mantled howler monkeys. See text for further discussion. ©Clara B. Jones

ways (access to limiting resources, thermoregulation). Cohen (1971) studied the statistical properties of frequency distributions of primate subgroups varying in size finding that, ceteris paribus, a zero-truncated binomial distribution provided the best fit where rate of replacement of individuals was >1. Thus, by definition, a subgroup must be an "open" aggregation of conspecifics with the potential to increase in size. Rannala and Brown (1994) theorized that expansion of subgroup size is expected to cease where individual numbers approximate some equilibrium value relative to prey size, dispersion, and quality. These authors also concluded that subgroup size may be inherently unstable where additional solitary individuals, or members of other demographic groups, decrease mean coefficient of subgroup relatedness (r).

Population structure has significant consequences for genes and the individuals that carry them and may be evident as subdivision into demographic subunits representing an evolutionary compromise among parameters yielding optimal inclusive fitness to individuals. The frequency distribution of group sizes will be a function of those phenomena leading individuals to join and to leave groups combined with the selection pressures on individual responses to these forces. Parameters determining modal group size in a population, thus, are ultimately expressed as adaptations of individuals to local conditions. Subgroup sizes of an aforementioned mantled howler monkey group (Jones 1980, 1995b) were sampled using *ad libitum* methods over an 18-month period. Only adults were counted (no. of adults in group = 18). Figure 4.3 displays subgroup counts and frequencies (mean = 4.46 ± 1.99, $N = 120$ subgroups). Coefficient of dispersion (CV) was 0.89, a repulsed (overdispersed) distribution with more observations at the center of the distribution than at the extremes and with variance smaller than one would expect by chance alone, indicative of optimal subgroup size.

Table 4.1 displays frequencies (*f*) of subgroups with different compositions. Males are identified by dominance rank (1, 2, and 3, highest to lowest rank). Mean ± standard deviation (SD) and coefficients of dispersion (CD) are shown for each subgroup composition (SGC). Female subgroups exhibited the lowest mean group size. Males belonged to subgroups with about equal frequency, and mean

Table 4.1 Subgroup composition (SGC), frequency (*f*), mean ± standard deviation (SD), and coefficient of dispersion (CD) of subgroups of one demographic group of mantled howler monkeys in seasonal tropical forest habitat

SGC	*f*	Mean ± SD	CD
Females	33	3.03 ± 1.24	0.51
2	28	4.71 ± 1.72	0.63
1	26	4.85 ± 1.43	0.42
3	24	5.17 ± 2.08	0.84
2, 3	5	7.60 ± 2.50	0.82
1, 3	4	2.75 ± 1.50	0.82

Behavioral tactics and strategies among howler monkeys are diverse, and patterns of food predation within and between *Alouatta* species can be explained, in part, by exploitation of limiting, mostly ephemeral, clumped, plant tissues and a niche breadth sensitive to local dispersion of food type and quality. Individuals of this genus, like many social mammals, frequently subgroup opportunistically for feeding, a foraging mode transitional between obligate or occasionally solitary feeders and vertebrate species appearing to subgroup as a characteristic, and, possibly, obligate, feature of diurnal foraging activity ("fission–fusion" societies: some primates, cetaceans, canids, and felids). In part because of their well-known habits, their wide geographical distribution, and their relatively generalized phenotypes, howler monkeys have the potential to serve as a model for primary consumers within and between vertebrate orders (Sect. 4.1)

group sizes of subgroups including a single male were, likewise, similar. Subgroups including two males reflected the dominance hierarchy whereby subgroups including second- and third-ranked males occurred with greater frequency than subgroups including first- and third-ranked males. Subgroups including the first- and second-ranked males were rare, consistent with results of a translocation experiment manipulating spatial relations among these three males in the field (Jones 1982). All coefficients of dispersion were repulsed. A one-tailed *t*-test applied to mean subgroup sizes with and without males revealed that subgroups including males were significantly larger, on average (≤ 0.001, $t = 7.88$, df = 118). It may benefit females to subgroup with males on some occasions (Rannala and Brown 1994), if males provide defense, if interindividual conflict is lower in subgroups including males, or if the latter subgroup composition provides females with greater access to preferred resources.

Giraldeau and Caraco (2000) treated social foraging theoretically, providing analyses of energetic "gains and losses" to individuals feeding with one or more conspecifics at the same time. These authors conceptualized social (concurrent) foraging as a series of discrete "decisions" measurable as variations in phenotypic traits and amenable to quantitative analysis. Giraldeau and Caraco (2000) pointed out that learning may influence variations in phenotypic traits displayed by individuals when foraging socially. Zink and Reeve (2005) generated "time-dependent functions of reproductive allocation" (productivity curves) to describe when it "makes sense" for individuals to remain in or to leave reproductive groups, an approach applicable to conditions in which individuals forage socially. Zink and Reeve (2005) identified factors likely to influence shapes of productivity curves (endogenous and exogenous "states"), pointing out that their models can be modified to describe variations in phenotypes (decisions, traits) influenced by learning, including differential probabilities of "success" (survival and reproduction). The quantitative approaches used by Giraldeau and Caraco (2000) and Zink and Reeve (2005) can be used to investigate parameters of adaptive decision-making and phenotypic variability where more than one conspecific exploit limiting resources concurrently ("contest competition").

Chapter 5
Learning May Generate Phenotypic Variability in Heterogeneous Regimes

"I find it difficult to believe that the neural basis of a complex set of actions should be built up out of chance mutations when in fact there is an apparently simpler route, by way of the process which Waddington (1953) has called 'genetic assimilation'."

Ewer (1968)

"Animals don't have to be intelligent; they simply must behave as if they are."

Wilson (1971)

"...animals choose the variable delay or the probabilistic reinforcer because the possibility of receiving a reinforcer after little or no delay outweighs the possibility that there must be a long delay before the next reinforcer is delivered."

Mazur (2004)

Keywords Learning • Deconstraint • Long-term potentiation (LTP) • Cooperative breeding • Saltzman's Marmoset Model • Reinforcement Reward

Compared to other vertebrates, the global success of terrestrial mammals, including social taxa (rodents, cetaceans, primates, carnivores: Coda), is often attributed to (1) generalized, (2) *totipotent* (reversible, flexible) traits (Jones 2005a, 2009), particularly characteristic of large-bodied mammals and many social species (among rodents, cetaceans, primates, carnivores, bats). These two properties may preadapt mammals for relatively efficient and effective thermoregulatory feedback in heterogeneous regimes, a topic highlighted by Sporn's (2011, Chap. 7) discussion of the mammalian brain's design economy. Classical conditioning, a form of associative learning, entails pairing a neutral stimulus (conditioned stimulus: CS) with a reflex

C.B. Jones, *Robustness, Plasticity, and Evolvability in Mammals: A Thermal Niche Approach*, SpringerBriefs in Evolutionary Biology, DOI 10.1007/978-1-4614-3885-4_5, © Clara B. Jones 2012

or other involuntary (autonomic) response (unconditioned stimulus: US) until the CS elicits the response (conditioned response: CR) originally elicited by the US (unconditioned response: UR). Classical conditioning exemplifies the interactive connectome of robust and labile responses combining and recombining motor patterns to form novel phenotypes.

Studying cooperatively breeding common marmosets (*Callithrix jacchus*), Saltzman (2003) suggested that classical conditioning explains the maintenance of ovulatory suppression in subordinates by dominant females, without, in this research condition, direct, tactile, or, even, targeted, directional interactions. In the *Callithrix* study, the dominant female harassed and intimidated (US) subordinate females (during group formation), effecting anovulation (UR) via a hypothesized, fear-induced, and "stress-mediated mechanism." Saltzman (2003) hypothesized that subordinates *associate* (pair) harassment and intimidation (US) with sensory (recurrent) cues (CS: neutral stimulus) from the dominant female (olfactory, visual, display), an olfactory mechanism demonstrated for *Drosophila* larvae by Eschbach et al. (2011) and Dulac and Torello (2003). After continuous pairings between US and CS, CS alone is sufficient to evoke and maintain anovulation (reproductive suppression: CR). Saltzman (2003) points out that the classically conditioned response may extinguish (cease to be elicited; Baum 1966) if events "disrupt" ("deconstrain") the CS: US connectome, a consequence likely to occur where the CR ceases to follow CS.

In Saltzman's (2003) model, the conditioned (acquired) response is physiological (biochemical: "ovulation suppression") rather than behavioral (motor), representing the mechanistic modification of a functional unit of the female's reproductive system with phenotypic effects upon which selection may act (suppression of reproduction, decreased competition with the dominant female, and a "switch" to "helper" role). Novel phenotypes induced by learning mechanisms may enhance an organism's inclusive fitness via condition- and rate-dependent interactions with the organism's biotic (including social) and abiotic (substrate, forest architecture) environments. An extinguished response acquired by learning persists as a neural "memory" that, when restimulated in similar conditions by the original US, will reappear with greater intensity than previously expressed. Most discussions of the potentially effective consequences of learning mechanisms fail to link learned responses to metabolism and thermal tolerances, to the organism's reproductive success, nor to mean population fitness.

Treatments of phenotypic variation generated by learning omit the concept of "reward" ("reinforcing stimulus") that, by definition, increases the likelihood that a response will be repeated. "Learning theory" ("behavioral psychology") recognizes two categories of reinforcement: primary (hunger, thirst, sex) and secondary (stimuli effective as reinforcement or reward because of their direct relationship with one or more primary reinforcer). During the developmental phases of behavioral psychology (also ethology), primary reinforcers were conceptualized as "drives" "motivating" animals (including humans) to respond (Wheeler 1928, Jones 2008). By the 1970s, however, these vague and controversial terms had been abandoned in favor of straightforward quantitative, including mathematical, descriptions of responses before and after occurrence of S relative to behavioral events before and after R.

In Saltzman's (2003) study, sex may be assumed as primary reinforcement and fear reduction (escaping harassment and intimidation via reproductive suppression) as secondary reinforcement. In reality, however, unless reinforcement is experimentally controlled and manipulated, the primary reinforcer may be difficult to determine with confidence. Elaboration of the aforementioned formulations is beyond the scope of the present review. However, the reader will note (1) sex is the only primary reinforcer for which sustained deprivation is not lethal, and (2) hunger and thirst are critical components of energetic processes, and sex (allocation) is an indirect effect of energetics, directly related to relative reproductive success.

Mazur's (2004; Baum and Rachlin 1969) "hyperbolic-decay model" clarifies the "reinforcement" ("reward") concept. This formulation expresses the value (V: strength, effectiveness) of a reward (to an individual receiving it) as a function of its temporal features. Thus,

$$V = A/1 + KD,$$

where V = value of an option that delivers a reinforcer after D seconds; A = amount of reward; K = a parameter describing how rapidly V decreases with increasing delay, D. Mazur (2004) provides explicit applications of his "hyperbolic-decay model," observing that (1) timescales vary with species, (2) results using human subjects are comparable to those for nonhuman subjects if reinforcements in the prior conditions are veridical (ecologically relevant), (3) relative to scale, V decreases rapidly; though, for *Homo sapiens*, V may take months → years to decrease. In the report under discussion, Mazur (2004) studied effects of variations in V in uncertain conditions ("risky choice"), and his research may be extended here to questions about differential norms of reaction where variations in effectiveness (V_e) of limiting proximate factors (food, water, mates, breeding sites: V_p; Figs. 3.1 and 3.2; Box 2.1) impact life-history strategies, reproductive allocation decisions, and, ultimately, mean fitness of populations, and V_p may have a neutral effect ($V_{=p}$), a deleterious effect (V_{-p}), or a beneficial effect (V_{+p}) on reproductive success (ultimate factor: V_u) and a population's mean fitness. Significantly, Vickery et al. (2011) have shown that, from "low-level sensory areas" to "high-level social reasoning," "distributed representation" of reward signals optimizes neuroethological functions in mammalian brains.

5.1   In Addition to Flexible Reaction Norms and Totipotency, What Other Characteristics Enhance Evolvability in Mammals?

Many mammal lineages (*Rattus*, *Homo sapiens*, didelphids) exploited thermal niches favoring "responsive switching" and "tracking," as well as opportunistic, facultative, rapid decision-making in response to uncertainty (variations in resource dispersion), risk (predation), and other environmental fluctuations (Jansen and Stumpf 2005; Kussell and Leibler 2005). Compared with other vertebrate classes, mammalian taxa exhibit an increased ratio of neocortex to total brain size (Armstrong 1983) comprised of modular, hierarchical networks of neural circuits ("connectome"), coordinating and controlling vital responses (most biophysical events, information processing, behavior) and capable of molecular and cellular reprogramming (recruitment, differentiation, repair: Jerison 1983; Matzel et al. 2003; Hsu et al. 2005; Eisenberg 1981). Although rates of gene expression levels in mammalian brains differ across lineages (Brawand et al. 2011), increased brain mass and body size occur at the same rate as a function of total energy available (Armstrong 1983; Hamilton et al. 2011). Modular organization (Sporns 2011, Chap. 7) is characterized by sensitivity to temporal cues, exposed to selection indirectly via phenotypically expressed biophysical traits. Relative brain enlargement is associated with a suite of traits (endothermy, diversification of clades, maternal care, primitive eusociality), energy-allocation programs (Fig. 3.1) incorporating "sensing," perception, learning, and memory, functions considered to enhance inclusive fitness in heterogeneous regimes ("bet-hedging," Jansen and Stumpf 2005; Kussell and Leibler 2005).

In fluctuating, unpredictable, or risky conditions, the mammalian cortex assesses likelihoods of reward and selfish gain (de Quervain et al. 2004; Wacongne et al. 2011), aptitudes with the potential to enhance inclusive fitness for individuals in heterogeneous regimes (Jansen and Stumpf 2005). Theoretical and empirical research have demonstrated the importance of ecological and social factors for the evolution of large brains ("the social brain hypothesis," conclusions consistent with reports linking thermal events in local ("patch") regimes to the evolution of biophysical and life-history characters) (Stearns and Koella 1986; Ricklefs and Wikelsky 2002; Hallgrímsson and Hall 2005). Studying marsupials and eutherians, Weisbecker and Goswami (2010) tested the hypothesis that relative brain size in mammals correlates highly with maternal allocation strategies and basal metabolic rates (BMRs). These authors demonstrated that, if Primates are excluded from analyses, relative brain size across therians was equivalent, contradicting the widely held proposition that large brain size is a necessary function of relatively higher BMRs. Weisbecker and Goswami (2010) showed, further, that relatively large brain size in marsupials and nonprimate eutherians is achieved via different trajectories, marsupials through "extended lactation," eutherians (sans Primates) by way of relatively longer periods of gestation afforded by a "true" placenta (Sect. 2.4).

These results confirm that the occurrence of similar traits in mammals is not sufficiently informative to justify a claim for similar origins and that differential evolutionary trajectories probably employ ancestral ("toolkit") traits relative to the abiotic and biotic selection pressures (and rates) of particular lineages. Conducting clever field experiments with female eastern gray kangaroos (*Macropus giganteus*) of reproductive age, Cripps et al. (2011) demonstrated high costs of lactation when gestating and nongestating adult females were compared. Though both of the aforementioned studies strongly suggest the metabolic costs of relatively large mammalian brains derived not only from cerebral structure and function (Weisbecker and Goswami 2010), but also from taxon-characteristic patterns of obligate maternal investment. These costs, however, by definition, must have been outweighed by reproductive benefits from endothermy (relatively higher reproductive rates compared to ectothermy, resistance to environmental perturbation, predation, and competition: McNab 2006). Mostly divergent reproductive allocation strategies by female birds and female mammals are a likely result of the oviparous habits and highly conserved phenotypes of the former class, characteristics possibly explaining evolutionary constraints in Monotremata.

Brain size: BMR is highly correlated across nonprimate eutherians but not across marsupials, and different life-history factors (different female energy-allocation strategies) explain these differences, supporting intuitive and quantitative conclusions reported by Eisenberg (1981) and Hamilton et al. (2011), respectively. Across Class Mammalia, Hamilton et al. (2011) demonstrated "universal" scaling of net production (litter mass at birth or weaning) despite independent evolutionary trajectories and variegated life-history strategies. According to these authors, their findings confirm that forces (pressures) on productivity follow the same metabolic algorithms across mammalian lineages sharing structural and functional similarities in biophysical organization (McNab 1980, 2006). Energy allocation strategies in mammals, then, derive from endothermy (McNab 1974, 1980,1986, 2006), a fundamental mammalian signature yielding differential reproductive costs and benefits to individuals' mean reaction norms (Hamilton et al. 2011).

Following the review by Jones (2008), in some vertebrate taxa, selection in fine grained (Box 2.1), "open" regimes, appears to explain "genetically coded," "open" behavioral programs, including learning processes. Learning (experience-dependent changes in behavior) is ubiquitous among complex (multicellular) organisms (Skinner 1935), possibly explained by highly conserved, regulatory toolkits comprised of basic molecular building blocks responsible for differential genome–phenome reorganization, recruitment, and expression over ecological and longer spatiotemporal scales (Skinner 1935; Kuraku and Kuratani 2011; Nilsonne et al. 2011). The particular tempos and modes responsible for the evolution of proteins implicated in learning and other mechanisms characterized by "constrained variation" are currently unspecified (Pasque et al. 2011). Nonetheless, recent empirical studies have documented temperature-dependent nucleosome binding, time-dependent "reconfiguration" (rearrangement, reorganization, reprogramming) of brain networks and their behavioral outputs, rapid rates of protein evolution (Vieira-Silva et al. 2011), associative learning by single cells in vitro, associative learning in *Drosophila* larvae, apparent rearrangement of several highly conserved gene networks, and highly conserved, indeed, "generic" mechanisms for Herrenstein's matching law (Loewenstein and Seung 2006), and Katsnelson et al. (2011) showed that, in theory, "learned strategy choice" can evolve by frequency-dependent selection in fluctuating regimes.

5.2   How Might Precursors of Learning Mechanisms Facilitate Evolvability by Learning?

Kirschner and Gerhart (1998) suggested that evolvable characters may originate as "trial-and-error" ("noisy," stochastic, dynamic, "excitatory") responses (Thorndike 1898). In a series of reports on his research using cats as subjects, Thorndike (1911) showed that "trial-and-error" motor patterns may initiate one or more learned, associative (stimulus–response) events resulting from recurrent interactions (associations) between environmental stimuli and an organism's biophysical outputs. "Excitatory" activities of biological organisms may induce "variability of response" (Hull 1934; Thorndike 1911), phenomena discussed as "latent activities" in contemporary scientific literature. Learned motor patterns, then, may derive from "latent activities" of neuromuscular mechanisms (Sect. 4.1), permitting novel, exploratory ("tinkering"), and hypervariable (extravagant and excessive behavioral repertoires) responses to dynamic abiotic and biotic features of the organism's (Budd 2006; Brakefield 2006; Lee et al. 2006; Tinbergen 1952) endogenous and exogenous states ("classical conditioning": Pavlov 1927, 1928). Novel phenotypes may originate also with "flooding," auto-shaping, tropisms, sensory biases, superstitions, and "prepared" dispositions as well as nonassociative learning processes, habituation, sensitization, and desensitization, to highlight a few examples.

Importantly, behavioral psychologists ("learning theorists") generally view optimal decisions and choices as temporal parameters (T: Baum and Rachlin 1969; Ferster and Skinner 1957; Skinner 1935; Mazur 2004), a topic of general import for studies of individuals and populations inhabiting fluctuating environments, particularly, the field's very large mammalian literature on conditional stimulus (signal, information) and response (receiver, experimental subject) interactions, including frequency, rate or uncertainty, intensity, duration, type (motor learning, imitation), accuracy ("matching"), and quality (gustation, reinforcement/reward). Additional research questions, methods, results, and conclusions in the field of behavioral psychology are likely to elucidate ways in which novel phenotypes arise as a result of environmental induction, particularly novel neuroethological variations without effective heritability. In studies on the consequences of classical conditioning, for example, Wasserman (1973) investigated use of heat as reinforcement, and Woods and Kulkosky (1976) demonstrated changes in blood glucose levels during classical conditioning (Chap. 3).

These reports directly link responses modified by associative learning (and, possibly, other variations in neuroethology) to thermosensory and thermoregulatory processes (Dell et al. 2011; Suggitt et al. 2011; Fig. 3.2), indirectly coupling "acquired" characteristics and local ("patch") conditions (Dell et al. 2011; Suggitt et al. 2011), relative inclusive fitness of individuals, and mean fitness of populations (Box 2.1, Figs. 2.3 and 4.2; Ricklefs and Wikelsky 2002; Schoener 2011). Wasserman's (1973) research suggests, further, that stimuli such as temperature vary along a reward–aversion continuum and that, in some conditions, a stimulus or stimulus array may be reinforcing, punishing, neutral, or some combination of these states. Variations in environmental factors inherent to treatments of the origins, mechanisms, maintenance, and consequences of genetic assimilation and evolvability (Kirschner and Gerhart 1998; West-Eberhard 2003) are fundamental to the behavioral psychology canon, a database widely applicable to the field of neuroethological plasticity. Related, as well, to research on robustness, plasticity, and evolvability, Woods and Kulkosky's (1976) research shows that the same database applies to variations in metabolic processes in response to variations in environmental stimuli, highlighting the need to integrate behavioral psychology into "the emerging field of eco-evolutionary dynamics" (Schoener 2011).

Supporting the aforementioned interpretations and extensions of learning theory, recent mammalian field research has demonstrated vital interactions among ambient temperature and plastic physiological/metabolic responses (*Rattus fuscipes*: Glanville and Seebacher 2010), among variations in "metabolic responsiveness [plasticity] to cold [temperature]" and "physical activity" (*Rattus fuscipes* behavior: Seebacher and Glanville 2010), as well as between variations in behavior and regulation of blood glucose levels (nectar-feeding bats: Kelm et al. 2011). In sum, the very large, quantitative, including experimental and mathematical, database on proximate causes and consequences of variations in learning and behavior will facilitate scientific understanding of the "physiology–life-history nexus" by specifying the effects of (controlled) variations in temporal parameters between variations in quantified individual responses (endogenous and phenotypic) and quantified variations in abiotic and biotic environmental stimuli. Of equal importance, behavioral psychology offers powerful methodological paradigms to ecological and developmental evolutionary biology for hypothetico-deductive research addressing endogenous and exogenous stressors and behavioral accommodation, including many questions related directly to evolvability ("novelty": Baum 1970, 1974).

5.3 How Limiting Is "Deconstraint" for the Evolution of Evolvability?

Kirschner and Gerhart (1998) pointed out that it remains unclear how novel expressions of cortical plasticity can arise from latent activities "without severely compromising the original activity," discussing several candidate mechanisms. One among these is long-term ("Hebbian") potentiation (LTP: the long-term enhancement of neural communication resulting from simultaneous, recurrent stimulation of synapses), a highly conserved vertebrate mechanism with the potential to explain the exhibition and maintenance of novel latent activities. Kirschner and Gerhart (1998) suggested that LTP may explain "behavioral accommodation" as a special case of plasticity whereby long-lasting cellular changes influence molecular, cellular, and, presumably, phenotypic expression. For most animal taxa, including mammals, a large literature documents LTP as a mechanism fundamental to learning and memory and as the primary mechanism of neural plasticity with potential to induce novel, heritable, and/or responses without effective heritability. Research on mammalian genome regulation is at an early stage, but a recent study by Brawand et al. (2011) provides insights into the functional variations associated with brain areas implicated in "motor neuron [behavioral] development and maintenance," including evidence that differential evolutionary rates of some "expression switches" account for variations in mammalian organ biology (brain, liver).

"Noise" or stochasticity is inherent to all forms of computational activity pro-ducing a baseline degree of error or inaccuracy of response (Cusack et al. 2011). Within constraints of critical error thresholds determined by an organism's thermal tolerances, nonlethal, noisy responses, such as "trial-and-error" action patterns, may enhance fitness (Katsnelson et al. 2011), possibly facilitating exploratory and "tinkering" properties (Kirschner and Gerhart 1998; Jones 2008). As potential inducers of learning mechanisms, "trial-and-error" activities may be particularly advantageous in microclimates whose conditions narrowly circumscribe informa-tion embedded in external signals (Laughlin and Sejnowski 2003). Under these conditions, cortical rearrangements induced by learning renders modified circuits more sensitive to sensory information. While environmental "noise" ("trial-and-error" events) is widely considered detrimental to system function, it may, under some conditions, enhance the sensitivity of a biological system (Katsnelson et al. 2011), possibly enhancing inclusive fitness. These ideas are in the early stages of investigation, particularly as they relate to learning mechanisms. Research on simi-lar questions at the protein level (RNA folding) highlights the importance of clearly differentiating organism and population levels of analysis (Wagner 2008), distinc-tions rarely made in behavioral psychology (or other areas of psychology) because investigators focus on proximate rather than ultimate causation. Phenotypic vari-ability resulting from, for example, social (observational) learning, modeling, or imitation is likely to increase at the individual, but not at the population level, since the former and related processes are likely to increase phenotypic similarity among members of a population.

Continuing to follow Jones' (2008) review, the mammalian brain's "latent" potential is also expressed as selective appropriation of molecules from one func-tion or set of functions to others (Kirschner and Gerhart 1998), events that may significantly influence epistasis, pleiotropy, and epigenetics. Candidate molecules potentially inducing novel neuroethological responses via LTP are PKC, MAP kinase, and PKA. Reviewing "responsive switching" by the amygdala, a conserved (vertebrate) temporal lobe structure implicated in emotional learning and memory, Maren (1999) showed that LTP mediates learning and memory in mammals. Similarly, anatomical, molecular, cellular, and neuroethological research has dem-onstrated the genetic and chemical control of amygdaloid LTP, including its facilita-tion and inhibition. Differentiated adult neurons have recruited molecular mechanisms used to regulate cell division and to perpetuate cell phenotype in order to perform one of its primary functions—long-term plasticity (Day and Sweatt 2011), another system demonstrating rearrangements, appropriations, and modifications of primitive architectures (Bassett et al. 2011). The aforementioned reports suggest that synapses detect and respond to events likely to "compromise" their "original activity," and it will be necessary to understand how and under what conditions one or more genetic switch mechanism enhances rather than prevents reprogramming (Abel et al. 1998).

Reviewing invertebrate literature, Abel et al. (1998) showed that epigenetic effects are implicated in consolidation (LTP) and stabilization of neurons related to cognitive-behavioral memories, providing support for an "epigenetic code" in the central nervous system (CNS) mediating plasticity, learning, and memory. These authors pointed out that, in the adult CNS of invertebrates, different epigenetic events underlying molecular and cellular memory consolidation may interact during memory formation (DNA × histone regulation). Extending Abel et al.'s (1998) work, invertebrate epigenomes are highly conserved for plasticity and memory formation (neural reorganization), and molecular homology (shared molecules and biochemistry) has been demonstrated for memory circuits (structure) and function (synaptic plasticity), results possibly applicable to mammals and to arguments in favor of generic "toolkit" programs (Coda).

Loewenstein and Seung (2006) showed by simulation that some learning programs are "remarkably conserved" across taxa, are dependent upon "generic," quantifiable rules governing neural plasticity, and exhibit present and retrospective "neural activities," including decision and choice, programmed by synaptic plasticity ("fluctuations in neural events"). Related to Loewenstein and Seung's (2006) studies, de Arcangelis and Herrmann (2010), using a topological approach, reported that all occurrences of "complex" (multifactorial) learning are, in theory, a function of "slow" synaptic changes, suggesting one temporal mechanism whereby system reorganization may occur cryptically, below thresholds required to induce inhibitory, repressive regulation. Working in concert, inhibitory processes are fundamental to functional "cortical circuitry," including maintenance of regulatory "homeostasis," permitting asynchronous, irregular network states essential to information transfer and processing (Vogels et al. 1990). Additional models of mammalian memory, learning, and behavior, including empirically based treatments (Somel et al. 2011), can be developed subsequent to comprehensive study of functional genomics (Brawand et al. 2011; Meyer et al. 2011, Houtkooper et al. 2011; Korb and Finkbeiner 2011; Burgess 2011) and applications of emerging methodologies (Kishida et al. 2011; Wang et al. 2009).

Chapter 6
Discussion: Stimulus ↔ Response ↔ Stimulus

> "We can define an arboreal folivore syndrome for tropical mammals. We can define an herbivorous fossorial syndrome, an insectivorous fossorial syndrome, a cursorial grazing syndrome, a cursorial browsing syndrome, and a cursorial carnivorous syndrome."
>
> Eisenberg (1981)

Keywords Thermal niche • Inclusive fitness • Evolvability • Responsive phenotypes • Alternative responses • Polyphenisms

6.1 Introduction

The scientific literature on robustness, plasticity, and evolvability is "massive." Each general topic is traceable to specialized reports on all manner of detail concerning every imaginable and unexpected event related to regulation and expression of spatiotemporally reliable as well as spatiotemporally variable reactions norms. For several reasons, robustness, plasticity, and, by implication, evolvability are not straightforward effects, primarily because phenotypic characteristics are direct functions of regulatory feedback processes sensitive to intra- and inter-component signaling and also because some of the most fundamental elements of biological systems are mediated by dynamic ("hypervariable") activity. In sum, my Springer Brief provides a broad treatment of phenotypic robustness and variability as outcomes of fluctuating and, sometimes, evolving, thermal niches interacting with thermal tolerances, differentially affecting individual survival and reproduction (decreased, increased, or unchanged inclusive fitness). Thermosensory surfaces (skin, eyes, ears) transducer and relay information to an organism's neurons whose ("Hebbian") synapses decide to respond or not to respond to features of the stimulus array. Decisions to terminate or proceed with transmission of neural information are

C.B. Jones, *Robustness, Plasticity, and Evolvability in Mammals: A Thermal Niche Approach*, SpringerBriefs in Evolutionary Biology, DOI 10.1007/978-1-4614-3885-4_6, © Clara B. Jones 2012

effected by inhibitory processes, producing recurrent and/or novel thermal tolerance responses (norms of reaction) determining the relative "fit" between a ("responsive") phenotype and a particular (dynamic) environmental state.

6.2   Investigating Genes and Phenotypes as Leaders as well as Followers Is an Inherently Reductionistic Program

Two key assumptions underlie most discussions of "phenotypic plasticity" (cf. West-Eberhard 2003) (1) phenotypes are "responsive" to environments and (2) "genes are followers, not leaders, in evolution." Although the subtleties of West-Eberhard's (2003) views remain under consideration (Fusco and Minelli 2010), it is clear at this point that several modifications to her verbal formulations are likely to be required. First, Chevin et al. (2010) and Lande (2009) stress the importance of quantitatively modeling aspects of "phenotypic plasticity." Jones (2005b) pointed out that differential costs and benefits, advantages and disadvantages of "phenotypic plasticity" are necessary functions of generation time relative to spatiotemporal patterns of environmental fluctuations. Using a graphical model, Jones (2005b, Fig. 1) showed, further, that benefits from phenotypic plasticity ultimately decline due to increased error or inaccuracy of responses with increases in environmental heterogeneity, and in unpredictable, stochastic regimes, phenotypic robustness will be favored. Following Schoener (1971), the previous author noted also that, due to metabolic differences, relative interests to individuals from "phenotypic plasticity" will depend on sex, age, and, possibly, other factors (matriline, "maternal effects," patriline, "behavioral syndromes," "early experience"), subject areas directly related to the possibility that suites of traits might characterize thermal tolerance development and variability, in particular, and patterns of thermal niche exploitation, in general.

Details of the cost and benefit curves schematized in Fig. 2.2 require dissection and diagnosis. For example, precise shapes of cost and benefit curves cannot be specified without knowledge of generation time (Box 2.1) and, perhaps, other parameters (female group size, age, intensity of sexual selection). Shapes of the curves for adult males and females in similar conditions are likely to differ (Austad 2006; Schoener 1971) because males are expected to be "time minimizers," and the more "plastic" (ancestral: Nijhout 2003a; Agrawal 2010) sex, females, "energy maximizers," and the more "robust" (derived: Libert et al. 2010; Lerner 1954; Jones 2005a) sex, leading to the inference that environmental heterogeneity affects males and females differently, effecting sexual dimorphism in access to resources, competitive regimes, sociosexual organization, and variability of lifetime reproductive success. *Caveats* obtain for hypotheses derived from differential energy investment in zygotes ("The Bateman Effect": males, small, plentiful, and cheap; females, large, few, and costly, Trivers 1972) because sexual dimorphism in reproductive strategies may result from (1) sexual selection stronger in males via male–male competition effecting higher variance in reproductive success while selection on females favors fecundity (Agrawal 2010); (2) reproductive benefits from parasitizing females may be the ancestral state for the latter sex ("sexual conflict": Rice 2000) rather than "The Bateman Effect," per se.

What information do we need for an approximate representation of cost and benefit curves for each (adult) sex? Compared to males in the same population, mammalian females assume a greater initial energetic investment in gametes since eggs are larger and more metabolically expensive to produce than sperm, and lactating females bear mammary glands producing nutrient-rich milk used to nourish offspring, a feature obligating them to offspring care, subordinate status, and "helper" roles. Mammaries enhance effects of anisogamy, with the potential to stress a female's "fitness budget" ("free energy" and allocation strategies), highlighting fundamental energetic distinctions between sexually mature males and females. Ceteris paribus, resting metabolic rate of (adult) female mammals is lower than that of (adult) males (Arciero et al. 1993), phenomena explained by higher ratios of body fat to total muscle mass rather than absolute body size differences between the sexes. Sexual dimorphism in zygotes, including phenomena following from it (alternative reproductive tactics and strategies: Jones and Agoramoorthy 2003), is the essential vertebrate polymorphism. Recent research suggests that testosterone may regulate not only male traits, but also life-history evolution for vertebrates as a whole (Hau 2007), possibly resolving a question addressed in empirical studies conducted by Weiss et al. (2006): Do sex-specific genetics explain particular traits or more general phenomena?

6.3   Measuring Variations in Thermal Tolerance and Relative Reproductive Success

Empirical research and field experiments on differential reproductive success of females is also indicated. Studying black howler monkeys (*A. pigra*) in Belize, Horwich et al. (2001) conducted an indirect test of female survivorship and/or fecundity by calculating effects of female group size on two female parameters. Following scientific literature describing the consequences of group size for adult females, the previous authors argued that increased group size may have no effect, a positive effect, or a negative effect on female survivorship or fecundity. Each of the possible outcomes reflected different causal mechanisms, and Horwich et al. (2001, Table 1) utilized the slope of the least squares regression line (b_{yx}) for female group size and number of immatures per group as proxies for mean female (thermal) tolerance curves. Tolerance curves (b_{yx}) could manifest as one of three phenotypic varieties: a slope ≈1 implies density-independent conditions in which female reproductive success (RS) is not expected to be a direct function of limiting food resources, for example, where deforestation, low population density, or other conditions (rapid rate of increase) induce unstable population dynamics. A slope >1 implies that there are advantages to living in groups (predator detection, efficient access to clumped resources), and female RS should increase with increasing group size, up to asymptote when density dependence (exhaustion of limiting resources) obtains (Fig. 2.2). A $b_{yx} < 1$ implies density dependence whereby females' thermal tolerances and reaction norms undergo significant stress from resource (food) competition, with deleterious effects on females' "fitness budgets" (Parker 1974). Selection is hypothesized to act on b_{yx}, and the aforementioned study demonstrated that slopes may vary between years, although the Horwich et al. (2001) study could not resolve the relative effects of micro- and macroclimate. In the heterogeneous regimes occupied by the study population, most of the yearly surveys (9 of 12, binomial test, $p = 0.0048$) exhibited slopes <1. Chevin et al. (2010) explore ways to explore similar phenomena with predictive models. While slopes of least squares regression lines can be employed as a proxy for a population's capacity to respond to environmental conditions (evolvability), more sensitive methods need to be developed incorporating data on variations in mortality, intrinsic rates of increase, and differential effects of micro- and macroclimates, providing estimates of selection intensity and rates.

6.4 Stress Induced by Fluctuating Abiotic and Biotic Variables May Induce Alternative Allocation Strategies, Including Social Behavior

Close proximity among conspecifics increases interaction rates, providing opportunities for interindividual exploitation. Studying parasitic birds, Jamieson et al. (2000) observed that social parasitism is most likely to be expressed in temporally and spatially heterogeneous regimes. Later research by Savolainen and Vespalainen (2003) concluded that polygyny is a prerequisite for intraspecific social parasites, often kin ("Emery's Rule"). Mammals are an excellent group for tests of exploitation theories because of their highly variegated behavior and sociosexual strategies.

Bondurianksy's (2011) functional logic, combined with ideas advanced by other authors (Rutherford 2000), exemplifies current research on the "responsive phenotype," including the induction of novel responses by social competition in specific ecological regimes (Allee 1926). A simple modeling exercise highlights the importance of socially selected parasitism for the activation, generation, and maintenance of phenotypic variation. For several reasons (initial energetic advantage, primary parasitic role derived from evolution of sex), males are usually assumed to initiate parasitic associations ("arms races": time-delayed feedback; despotisms) with subordinates, particularly, females (host class or state: Rice 2000). In the present example, consider a female mammal (biostate) parasitizing (stressing) the time budget of a male host in (delayed) response to the imposition of reproductive costs by males. Following May and Anderson (1990, in Moore 2002), Moore pointed out that fitness of a parasite (here, a reproductive female mammal) can be measured as reproductive rate (R_0, Sect. 3.3), a density-dependent value. May and Anderson's equation linking parasite transmission (female signals and displays, information transmission, communication; Sect. 4.3) to a parasite's influence on its host (deleterious effects on variance in male reproductive success) is related to virulence (rate of deleterious effects) by way of a measure of cost to host fitness (increased intensity of sexual selection on males). May and Anderson's equation can be modified for reproductive parasitism such that

$$R_{o} = y(N)/(a+b+v),$$

where y is transmission, N is host population density, a is rate of host cost, b is rate of host cost from all but virulence (opportunity costs), and v is recovery rate, a host's ability to completely or partially escape from or compensate for deleterious effects of a parasite, for example, by increasing future reproductive rate or switching to less time-intense (energetically costly, relatively more deleterious to reproductive success) mate monopolization strategy.

In the case described by Jones (1997c; Sect. 4.3), *Alouatta palliata* females in estrus may parasitize males (hosts) with a "rear-present" posture (multimodal visual–chemical display) by inducing males to follow to a preferred feeding source (flowering tree), subsequently defended by males for the duration of the female's feeding bout. After feeding, a female may or may not "decide" to copulate with her host, a response probably dependent upon male quality, latency since previous servicing, cycle stage, reproductive state, and other factors. By feeding and/or copulating, a female howler monkey in the described condition may increase her reproductive rate by, for instance, increasing her nutritional status, decreasing her interbirth interval (IBI), escaping harassment and/or competition from other males or females. Decreased IBI (increased R_{o}: increased reproductive success, increased rate of energy intake, increased rate of conversion of energy into offspring) is a function of signal effectiveness and virulence. As Moore (2002) pointed out, R_{o} increases as a decreases when virulence, transmission, and recovery rate are independent (when benefits of coevolution [sexual conflict, "arms race"] decrease: resource predictability, "evenness," or quality increase beyond some threshold value of reproductive benefit to the parasite). Under these conditions, the parasite (female) should evolve toward a (more or less) harmless state (decreased resistance to copulation, decreased "choosiness," mate selectivity (?), multiple mating by females (?), "monogamy," decreased "female emancipation," decreased benefits from ritualized signals and displays?). In such conditions, the potential for (and benefits of) female exploitation of males should be minimized (*Brachyteles* ?). Where virulence, transmission, and/or recovery rate are related, however, reproductive parasitism should be favored, and the degree of virulence should be determined by the relative degree of benefit to the social parasite, *ceteris paribus*.

Moore's (2002) verbal expansion of the 1990 model advanced by May and Anderson combined with the present application of the 1990 model to reproductive parasitism (Jones 1997c, 2005c) includes alternative definitions of the model's parameters. Applications suggested in this treatment relate resource dispersion, differential costs, and benefits from female (or, male, where males parasitize females) exploitation of males, directional (deleterious, advantageous, neutral) effects of reproductive parasites on their hosts, etc., have the potential to explain many aspects of and variations in social organization among mammals and, possibly, other vertebrates. For example, using a comprehensive review of primates as a baseline reference, the aforementioned model revised for social parasitism (y, N, a, b, v) may facilitate comparative field research on mammals (Chaps. 3 and 4), for example, the polygynandrous, *Brachyteles arachnoides* (very low rates of overt conflict or aggression within and between sexes, mostly evenly dispersed folivorous diet, male philopatry, female dispersal) with *A. palliata* (relatively low rates of overt conflict or aggression within and between sexes, aggression somewhat lower in males, not particularly affiliative, low to moderate dependence on leaves, diet comprised of spatiotemporally "patchy" fruit, flowers, new leaves, bisexual dispersal) with *Papio cynocephalus* (relatively high rates of aggression within and between sexes, particularly males, combined with relatively pronounced affiliative features among females, particularly, kin, relatively affiliative associations among nonkin, extremely fluctuating and stressful environments, unpredictable and relatively low-quality food, male dispersal, female philopatry). The aforementioned subjects as well as related questions and propositions wait theoretical (verbal and, especially, quantitative), modeling, as well as naturalistic description and controlled experimentation under laboratory and field conditions. Physiological capacitance may obtain to animals for enhancement of thermal tolerances. Evolvability fundamentally concerns thermal tolerances since evolvability ultimately entails effective and differential capacities to mitigate stress, mediated by and functions of the regulatory feedback events schematized in Fig. 3.1 (Crook 1965; Emlen and Oring 1977; Allee 1926).

Chapter 7
Synopsis

> *"Every increase in the diversity of the environment during the history of the world has resulted in a veritable burst of speciation. This is particularly easily demonstrated for changes in the biotic environment. The rise of the vertebrates was followed by a spectacular development of trematodes, cestodes, and other vertebrate parasites."*
>
> Mayr (1976)

Keywords Consumer-resource interactions • Relative reproductive success • Interspecific convergencies • Competition for limiting resources • Conserved intercellular communication • Information-transfer pathways

Any change in consumer-resource interactions may have positive, negative, or neutral effects on relative reproduction of consumers and resources as well as mean fitness of populations. At all levels of organization, individuals are characterized by "core" patterns, processes, and units (switches, circuits, protein regulation, neurotransmitter substances, motor patterns, recruitment), each of which is *differentially responsive* to abiotic and biotic (including social) stress. Some taxa appear to be preadapted to heterogeneous, even, near-lethal conditions (folivores), exhibiting significant thermal tolerances and resistance to thermal stress (desiccation). Flexibility as well as integration of system units is regulated by communication ("crosstalk") and regulatory feedback between thermosensors, mitochondria, and transcription factors demonstrating "remarkable versatility" for many types of (robust and plastic) reorganization at the molecular level. Gene regulation is amenable to epigenetic effects, particularly ecological (microclimate) factors (variations in temperature, food), information that is stored in neural "memory," influencing future molecular and cellular, including synaptic ("Hebbian"), events. Relative to

C.B. Jones, *Robustness, Plasticity, and Evolvability in Mammals: A Thermal Niche Approach*, SpringerBriefs in Evolutionary Biology, DOI 10.1007/978-1-4614-3885-4_7,
© Clara B. Jones 2012

system properties and tolerances, as well as spatiotemporal variations in limiting (functional ecological) resources (food items, nutrients, mates, breeding, conspecifics), energy is allocated to maintenance activities (system repair, homeostasis), survival (responses to parasitism, predation, interference, stress), and reproduction. At a given time, size of energy stores is limited ("elastic limit") and limiting, but, in mammals, female thermal stores ("fat") are proportionately larger compared to males in the same population. Some mammalian orders (rodents, bats, primates) include a number of species capable of storing energy in food pouches, and it may be of note that these orders include, as well, many social species. Several researchers highlight the paucity of field data on mammalian norms of reaction across ecological gradients, pointing out that, although much is known about separate stages from mammal genotypes to phenotypes, little is known about how the stages are coordinated and controlled from one level of system organization to another.

Literature on thermal tolerance evolvability in mammals does not usually weigh the significance of variations in microclimate for relative reproductive success, including inclusive fitness. Ceteris paribus, nondamaging responses (cooperation, altruism) should stress male allocation strategies because they are costly in T. Thus, where demographic (population structure, within and between "group" competition) or other (ecological, phylogenetic) constraints favor group formation (Sect. 3.3), an immediate challenge obtains to determine what energetic benefits of T-intense strategies outweigh costs (increased likelihood of incest and male–male competition) for males. At least for purposes of sex, the ability of males to monopolize females will be a direct function of ("map" onto) females' spatiotemporal dispersion (Emlen and Oring 1977), and, in mammals (as in social insects: Coda), varying degrees of sexual segregation are the norm, whether or not males reside in bisexual "social" units. Mate and offspring solicitation ("helping") by reproductively active male mammals is absent, uncommon, facultative, imposed, coerced, forced, transient, and/or the result of persuasion, and the highest "grades" of mammalian sociality are, in effect, female phenomena (cooperative breeding, eusociality), similar to social insects (Coda). Exceptions may exist (where males are severely constrained ecologically: bisexual co-residence, including "monogamy"; exploitation of male T-budgets by females; "female dominance"), and scientific analyses of these cases may await expurgation of anthropomorphic and other obfuscating terminology from animal behavior ("monogamy," "contact comfort," "empathy," "mourning," "bonds," "relationships").

What "calculus" of benefits and costs might favor sexual integration rather than sexual segregation among adult mammals in the same population? Hamilton's (1964) rule, $rb - c > 0$, predicts that (T-limited) "social" behavior (behavior benefiting a recipient's reproductive success more than an actor's) *toward relatives* (progeny and others bearing some threshold proportion of actor's alleles) is likely to be favored where c is an altruist's reproductive cost (direct reproductive costs), b, a recipient's reproductive benefit, and r, the coefficient of relatedness (inclusive fitness). It is more realistic to think of b and c as a ratio (Chap. 5). Theoretical evaluations of differential costs and benefits to inclusive fitness of male mammals are rare, and even rarer when the search is narrowed to studies viewing "social" behavior as a functional response to costs and constraints occasioned by thermal stress (abiotic: temperature; biotic: ecological). This outlook and set of assumptions permits interpolation of West et al. (2002) to differential costs and benefits of "social" behavior to male mammals, and, in theory, r, b, c can be estimated for males in a population, parameters that might, as well, permit estimation of optimal strategies given sufficient data about situational factors. Detailing and discussing the suggested interpolation are beyond the scope of the present review, and estimation of b and c across time and space will be a "daunting" task.

Several topics requiring theoretical, empirical, including experimental, attention have not been targeted with sufficient detail in my Springer Brief: in addition to eusocial behavior (Wheeler 1928), pleiotropy, epistasis, modularity, sexual selection, levels of selection (Wright 1949), reproductive skew relative to evolvability, critical and sensitive periods, immune responses to variations in temperature, as well as constraints imposed by robustness on resilience. A research program dedicated to the investigation of methodological approaches to "development and evolution of adaptive polyphenisms" needs prioritization. For example, if species with conserved molecular traits and minimal capacity to "acquire" habits can be identified, these might be employed as a "control group" for related species exhibiting ranges of phenotypic variability. Another approach would be to arrange a group of species along a continuum of phenotypic variability for study. These and numerous other nascent hypotheses amenable to scientific investigation have been addressed in the present review with potential to generate ideas and research.

7.1 Coda

As noted above, Mayr (1976) concluded that most animals are "specialized for a very narrow niche," a derived state, and it can be inferred from the latter proposition that there are an unimaginably large number of constricted, inflexible, and insular biophysical spaces ("lightless caves," "the interstices between sand grains"). Remarkably, in varying degrees, virtually all mammals retain features of animals' primitive, generalized state, despite their being characterized, as well, as the most diverse, widely distributed, and globally successful members of the Class. Generalized body plans combined with highly differentiated ecological (niche) adaptations make mammals the ranking control group for comparative vertebrate studies (Sect. 4.1; Chap. 3), from molecular to sociosexual levels and higher, including intra- and inter-specific competition, consistent with Eisenberg's (1981) rationales for highlighting marsupials as the candidate Order to serve as a "control group" for the Class.

Wilson (1971) considered social insects exemplary case studies for the investigation of biological organization because of (1) their use of specialized social traits "to solve ecological problems ordinarily dealt with by single organisms," (2) their terrestrial dominance, (3) their primacy as predators of invertebrate prey, (4) biomass and energy consumption surpassing vertebrates "in most terrestrial habitats," particularly, the tropics (Sect. 3.3). Noting that "extreme specialization" is virtually absent (*sic*) in plants, Mayr (1976) stated that the characteristic flourishes in Class Insecta explain "their prodigious rate of speciation." Importantly, for a study of contrasts between insects and mammals, the latter class is characterized by generalist characteristics yet, compared to other vertebrates, is widely dispersed and relatively speciose (Chap. 3).

Eisenberg (1981, pp. 438–445) attempted to trace intertaxon transitions from relatively undifferentiated to highly differentiated phenotypic and social programs, strongly suggesting that a comparative biology of insects and mammals would be productive. Insects and mammals are characterized by sexual segregation and have taken radically different paths to sociality, two of many comparisons waiting investigation. Apparent interspecific convergences have been noted between insects and social mammals (social parasitism: Jones (2005c), Table 1; dominance hierarchies, Jones (1980): honey bees, *Apis mellifera*; Hanuman langurs, *Semnopithecus entellus*, formerly, *Presbytis entellus*; mantled howler monkeys, *Alouatta palliata*). Honey bees feed on pollen and nectar, and both primate species are folivores, raising the possibility that convergent regulatory processes, including their precursors, will be identified, possibly, CYP2B, an enzyme necessary for "biotransformation" (blocking) of plant secondary compounds, in particular, terpenes (Haley et al. 2007). Such data might demonstrate a link between convergent genomes, metabolic events, and social phenotypes (in the present examples, structural features of dominance hierarchies and differential "resource-holding power"; Woodard et al. 2011), another connection to competition for limiting resources, in particular, food (energetics of folivores: Krause et al. 2011).

A research program integrating the topics addressed in this Coda, as well as other observations related to the comparative biology of insects and mammals, seems realistic based on research demonstrating convergent regulatory events, components, and circuits because: "complex intercellular communication is highly conserved" (Heidel et al. 2011); "the insulin signaling network is a key player that integrates… metabolism and the response to nutrition" (Nijhout 2003b); terrestrial vertebrates exhibit conserved "mitochondrial membrane proteins" (Kitazoe et al. 2011); "mammalian cells reveal conserved, functional protein turnover" (Cambridge et al. 2011); and downregulation ("deconstraint": Kirschner and Gerhart 1998, p. 8426) of the insect (*Drosophila*) "heat-shock" protein, Hsp90, "may be a molecular mechanism for promoting evolvability" (Rutherford 2000). Following Kirschner and Gerhart (1998), the components of all of these processes are characterized by "information-transfer [regulatory] pathways" extending back to "metazoan evolution." Reports summarized in the present review highlight the potential for general laws of "controlled variation," aspects of which are in development by researchers studying social insects, particularly in the laboratories of Gene Robinson, as well as the team, Joan Strassmann and David Queller.

References

Abel T, Martin KC, Bartsch D, Kandel ER (1998) Memory suppressor genes: inhibitory constraints on the storage of long-term memory. Science 279:338–341

Abrahms TW, Kandel ER (1988) Is contiguity-detection in classical conditioning a system or a cellular property? Learning in *Aplysia* suggests a possible molecular site. Trends Neurosci 11:128–135

Adams DK, Sewell MA, Angerer RC, Angerer LM (2011) Rapid adaptation to food availability by a dopamine-mediated morphogenetic response. Nat Commun 2:592. doi:10.1038/ncomms1603

Agrawal AF (2010) Are males the more 'sensitive' sex? Heredity 107:20–21

Allee WC (1926) Distribution of animals in a tropical rainforest with relation to environmental factors. Ecology 7:445–468

Allee WC (1931) Animal ecology of the tropical rainforest. Ohio State Univ Bull 36:95–102

Allen GR, Simmons LW (1996) Coercive mating, fluctuating asymmetry and male success in the dung fly *Sepsis cynipsea*. Anim Behav 52:737–741

Arciero PJ, Goran MI, Poehlman ET (1993) Resting metabolic rate is lower in women than in men. J Appl Physiol 75:2514–2520

Ariely D (2010) Predictably irrational. Harper Perrenial, New York

Armstrong E (1983) Relative brain size and metabolism in mammals. Science 220:1302–1304

Austad SN (2006) Why women live longer than men: sex differences in longevity. Genet Med 3:79–92

Aziz A, Liu Q-C, Dilworth FJ (2010) Regulating a master regulator. Epigenetics 5:691–695

Badyaev AV (2005) Stress-induced variation in evolution: from behavioural plasticity to genetic assimilation. Proc R Soc Lond B 272:877–886

Bassett DS, Wymbs NF, Porter MA, Mucha PJ, Carlson JM, Grafton ST (2011) Dynamic reconfiguration of human brain networks during learning. Proc Natl Acad Sci USA. doi:10.1073/pnas.108985108

Bates DG, Cosentino C (2011) Validation and invalidation of systems biology models using robustness analysis. IET Syst Biol 5:229–244

Baum M (1966) Rapid extinction of an avoidance response following a period of response prevention in the avoidance apparatus. Psychol Rep 18:59–64

Baum M (1970) Extinction of avoidance responding through response prevention (flooding). Psychol Bull 74:276–284

Baum M (1974) On two types of deviation from the matching law: bias and undermatching. J Exp Anal Behav 22:231–242

Baum M, Rachlin HC (1969) Choice as time allocation. J Exp Anal Behav 12:861–874

Baythavong BS (2011) Linking the spatial scale of environmental variation and the evolution of phenotypic plasticity in fine-grained environments. Am Nat 178:75–87. doi:10:10.1086/660281

Bell AM, Robinson GE (2011) Behavior and the dynamic genome. Science 332:1161–1162

Belovsky GE (1984) Herbivore optimal foraging: a comparative test of three models. Am Nat 124:97–115

Benoit-Bird KJ, Würsig B, McFadden CJ (2004) Dusky dolphin (*Lagenorhynchus obscurus*) foraging in two different habitats: active acoustic detection of dolphins and their prey. Mar Mamm Sci 20:215–231

Bicca-Marques JC, Prates HM, Rodrigues F, de Aguiar C, Jones CB (2008) Survey of *Alouatta caraya*, the black-and-gold howler monkey, and *Alouatta guariba clamitans*, the brown howler monkey, in a contact zone, State of Rio Grande do Sul, Brazil: evidence for hybridization. Primates 49:246–252

Blackburn DG, Flemming AF (2011) Invasive implantation and intimate placental associations in a placentotrophic African lizard, *Trachylepis ivensi* (Scincidae). J Morphol. doi:10.1002/jmor.11011

Bodmer RE (1990) Fruit patch size and frugivory in the lowland tapir (*Tapirus terrestris*). J Zool 222:121–128

Bonduriansky R (2011) Sexual selection and conflict as engines of ecological diversification. Am Nat 178. doi:10.1086/662665

Bradbury JW, Vehrencamp SL (1998) Principles of animal communication. Sinauer Associates, Sunderland, MA

Braendle C, Flatt T (2006) A role for genetic accommodation in evolution. BioEssays 28:868–873

Brakefield PM (2006) Evo-devo and constraints on selection. Trends Ecol Evol 21:362–368

Brawand D, Soumillon M, Necsulea A, Julien P, Csárdi G, Harrigan P, Weier M, Liechti A et al (2011) The evolution of gene expression levels in mammalian organs. Nature 478:343–348

Bubliy OA, Kristensen TN, Kellermann V, Loeschcke V (2011) Plastic responses to four environmental stresses and cross-resistance in a laboratory population of *Drosophila melanogaster*. Funct Ecol. doi:10.1111/j.1365-2435.2011.01928.x

Budd GE (2006) On the origin and evolution of major morphological characters. Biol Rev Camb Philos Soc 81:609–628

Burgess DJ (2011) Comparative genomics: mammalian alignments reveal human functional elements. Nat Rev Genet 12:806–807

Burton T, Killen SS, Armstrong JD, Metcalf NB (2011) What causes intraspecific variation in resting metabolic rate and what are its ecological consequences? Proc R Soc Lond B. doi:10.1098/rspb.2011.1778

Cambridge SB, Gnad F, Nguyen C, Bermejo JL, Krüger M, Mann M (2011) Systems-wide proteomic analysis in mammalian cells reveals conserved, functional protein turnover. J Proteome Res. doi:dx.doi.org/10.1021/pr101183k

Campbell C, Fuentes A, MacKinnon K, Bearder S, Caramaschi FP, Nascimento FF, Cerqueira R, Bonvicino CR (2011) Genetic diversity of wild populations of the grey short-tailed opossum, *Monodelphis domestica* (Didelphimorphia: Didelphidae), in Brazilian landscapes. Biol J Linn Soc 104:251–263

Cannon WB (1932) The wisdom of the body. W.W. Norton, New York, NY

Caramaschi FP, Nascimento EF, Cerqueira R, Bonvicino CR (2011) Genetic diversity of wild populations of the grey short-tailed opossum, *Monodelphis domestica* (Didelphimorphia: Didelphidae), in Brazilian landscapes. Biol J Linn Soc 104:251–263

Casci T (2011) Quantitative traits: gender inequalities. Nat Rev Genet 12:667

Castellana S, Vicario S, Saccone C (2011) Evolutionary patterns of the mitochondrial genome in metazoa: exploring the role of mutation and selection in mitochondrial protein-coding genes. Genome Biol Evol 3:1067–1079

Centola D, González-Avella JC, Equíluz V, San Miguel M (2011) Homophily, cultural drift, and the co-evolution of cultural groups. J Conflict Resolut 55:905–929

Chandrasekaran S, Ament SA, Eddy JA, Rodriguez-Zas SL, Schatz BR, Price ND, Robinson GE (2011) Behavior-specific changes in transcriptional modules lead to distinct and predictable neurogenomic states. Proc Natl Acad Sci USA 108:18020–18025

Charlton BD, Ellis WAH, McKinnon AJ, Brumm J, Nilsson K, Fitch WT (2011) Perception of male caller identity in koalas (*Phascolarctos cinereus*): acoustic analysis and playback experiments. PLoS One 6:e20329

Chevin L-M, Lande R, Mace GM (2010) Adaptation, plasticity, and extinction in a changing environment: towards a predictive theory. PLoS Biol 8:e10000357

Christie MR, Marine ML, French RA, Blouin MS (2011) Genetic adaptation to captivity can occur in a single generation. Proc Natl Acad Sci USA 109(1):238–242. http://www.pnas.org/cgi/doi/10.1073/pnas.1111073109

Cohen JE (1971) Casual groups of monkeys and men. Harvard University Press, Cambridge

Conradt L (1998) Could asynchrony in activity between the sexes cause intersexual social segregation in ruminants? Proc R Soc Lond B 265:1359–1363

Cooper N, Purvis A (2010) Body size evolution in mammals: complexity in tempo and mode. Am Nat 175:727–38. doi:10.1086/652466

Cooper N, Freckleton RP, Jetz W (2011) Phylogenetic conservatism of environmental niches in mammals. Proc R Soc Lond B 278:2384–2391

Coutinho-Abreu IV, Mukbel R, Hanafi HA, Fawaz EY, El Hossary SS, Wadsworth M, Stayback G, Pitts DA, Abo-Shehada M, Hoel DF, Kamhawi S, Ramalho-Ortigão M, McDowell MA (2011) Ecological variation in gene expression vectors. BMC Ecol. http://www.biomedcentral.com/1472-6785/11/24. Accessed 6 Dec 2011

Cripps JK, Wilson ME, Elgar MA, Coulson G (2011) Experimental manipulation of fertility reveals potential lactation costs in a free-ranging marsupial. Biol Lett. doi:10.1098/rsbl.2011.0526

Crockett CM, Eisenberg JF (1987) Howlers: variations in group size and demography. In: Smuts BB, Cheney DL, Seyfarth RM, Wrangham RW, Struhsaker TT (eds) Primate societies. University of Chicago Press, Chicago

Crook JH (1965) The adaptive significance of avian social organization. Symp Zool Soc Lond 14:181–218

Crozier LG, Scheuerell MD, Zabel RW (2011) Using time series analysis to characterize evolutionary and plastic responses to environmental change: a case study of a shift toward earlier migration date in sockeye salmon. Am Nat 178:755–773

Cui Q, Purisima EO, Wang E (2009) Protein evolution on a human signaling network. BMC Syst Biol 3:21. doi:10.1186/1752-0509-3-21

Cusack BP, Arndt PF, Duret L, Crollius HR (2011) Preventing dangerous nonsense: selection for robustness to transcriptional error in human genes. PLoS Genet 7:e1002276

Dantzer B, Swanson EM (2011) Mediation of vertebrate life histories *via* insulin-like growth factor-1. Biol Rev. doi:10.1111/j.1469-185X.2011.00204.x

Day JJ, Sweatt JD (2011) Epigenetic mechanisms in cognition. Neuron 70:813–829

Dayan E, Cohen LG (2011) Neuroplasticity subserving motor skill learning. Neuron. doi:10.1016/j.neuron.2011.10.008

De Arcangelis L, Herrmann HJ (2010) Learning as a phenomenon occurring in a critical state. Proc Natl Acad Sci USA 107:3977–3981

De Jong G (1976) Selection always increases efficiency. Am Nat 110:1013–1027

De Quervain DJ, Fischbacher U, Treyer V (2004) The neural basis of altruistic punishment. Science 305:1246–1247

Dell AI, Pawar S, Savage VM (2011) Systematic variation in the temperature dependence of physiological and ecological traits. Proc Natl Acad Sci USA 108:10591–10596

Denver RJ (1997) Proximate mechanisms of phenotypic plasticity in amphibian metamorphosis. Am Zool 37:172–184

Dulac C, Torello AT (2003) Molecular detection of pheromone signals in mammals: from genes to behaviour. Nat Rev Neurosci 4:551–562

Dumont ER, Dávalos LM, Goldberg A, Santana SE, Rex K, Voigt CC (2011) Morphological innovation, diversification and invasion of a new adaptive zone. Proc R Soc Lond B. doi:10.1098/rspb.2011.2005

Dunn LC (1921) Unit character variation in rodents. J Mammal 2:125–140

Durban JW, Pitman RL (2011) Antarctic killer whales make rapid, round-trip movements to sub-tropical waters: evidence for physiological maintenance migrations? Biol Lett. doi:10.1098/rsbl.2011.0875

Eberhard WG (1985) Sexual selection and animal genitalia. Harvard University Press, Cambridge

Eibl-Eibesfeldt I (1989) Human ethology. Aldine de Gruyter, New York

Eisenberg JF (1981) The mammalian radiations: an analysis of trends in evolution, adaptation, and behavior. University of Chicago Press, Chicago

Elliot MC, Crespi BJ (2006) Placental invasiveness mediates the evolution of hybrid inviability in mammals. Am Nat 168:114–120

Elton C (1936) Animal ecology. Macmillan, New York

Emlen JM (1973) Ecology: an evolutionary approach. Addison-Wesley, Reading, MA

Emlen ST, Oring LW (1977) Ecology, sexual selection, and the evolution of mating systems. Science 197:215–223

Enquist M, Hurd PL, Ghirlanda S (2010) Signaling. In: Westneat DE, Fox CW (eds) Evolutionary behavioral ecology. Oxford University Press, Oxford

Erwin DH, Laflamme M, Tweedt SM, Sperling EA, Pisani D, Peterson KJ (2011) The Cambrian Conundrum: early divergence and later ecological success in the early history of animals. Science 334:1091–1097

Eschbach C, Cano C, Haberkern H (2011) Associative learning between odorants and mecha-nosensory punishment in larval *Drosophila*. J Exp Biol 214:3897–3905

Espinosa-Soto C, Martin OC, Wagner A (2011) Phenotypic robustness can increase phenotypic variability after nongenetic perturbations in gene regulatory circuits. J Evol Biol 24. doi:10.1111/j.1420-9101.2011.02261.x

Estes R (1992) The behavior guide to African mammals including hoofed mammals, carnivores, and primates. University of Chicago Press, Chicago, IL

Evans AR, Jones D, Boyer AG, Brown JH, Costa DP, Ernest SKM, Fitzgerald EMG, Fortelius M, et al. (2012) The maximum rate of mammalian evolution. Proc Natl Acad Sci USA 109(11):4187–4190. http://www.pnas.org/cgi/doi/10.1073/pnas.1120774109

Ewer RF (1968) Ethology of mammals. Logos, London

Ferster CB, Skinner BF (1957) Schedules of reinforcement. Appleton-Century-Crofts, New York

Fetsch CR, Ponget A, DeAngelis GC, Angelaki DE (2011) Neural correlates of reliability-based cue weighting during multisensory integration. Nat Neurosci. doi:10.1038/NN.2983

Feldhamer GA, Drickamer LC, Vessey SH, Merritt JF (2004) Mammalogy, 2nd Edition. McGraw Hill, New York

Fleming TH, Breitwisch R, Whitesides GH (1987) Patterns of tropical vertebrate frugivore diversity. Annu Rev Ecol Syst 18:91–109

Flück M (2006) Functional, structural and molecular plasticity of mammalian skeletal muscle in response to exercise stimuli. J Exp Biol 209:2239–2248

Francis DD, Szegda K, Campbell G, Martin WD, Insel TR (2003) Epigenetic sources of behavioral differences in mice. Nat Neurosci 6:445–446

Frank SA (2011) Natural selection. II. Developmental variability and evolutionary rate. J Evol Biol. doi:10.1111/j.1420-9101.2011.02373.x

Frazier AE, Kiu C, Stojanovski D, Hoogenraad J, Ryan MT (2006) Mitochondrial morphology and distribution in mammalian cells. Biol Chem 387:1551–1558

Fusco G, Minelli A (2010) Phenotypic plasticity in development and evolution: facts and concepts. Philos Trans R Soc B 365:547–556

Gaillard JM, Yoccoz NG, Lebreton JD, Bonenfant C, Devillard S, Loison A, Pontier D, Allaine D (2003) Generation time: a reliable metric to measure life-history variation among mammalian populations. Am Nat 166:119–123

Gazzaniga MS (ed) (2004) The cognitive neurosciences (III). MIT Press, Cambridge, MA

Gellner G, McCann K (2012) Reconciling the omnivory-stability debate. Am Nat 179:22–37

Ghalambor CK, McKay JK, Carroll SP, Reznick DN (2007) Adaptive versus non-adaptive pheno-typic plasticity and the potential for contemporary adaptation in new environments. Funct Ecol 21:394–407

Gibb AC, Ashley-Ross MA, Pace CM, Long JH Jr (2011) Fish out of water: terrestrial jumping by fully aquatic fishes. J Exp Zool 313A. doi:10.1002/jez.711

Gifford RJ (2011) Viral evolution in deep time: lentiviruses and mammals. Trends Genet 28:89–100

Giraldeau L-A, Caraco T (2000) Social foraging theory. Princeton University Press, Princeton

Gittleman JL, Thompson SD (1988) Energy allocation in mammalian reproduction. Am Zool 28:863–875

Glander KE (1975) Habitat and resource utilization: an ecological view of social organization in mantled howling monkeys. Unpublished Ph.D. dissertation, University of Chicago, Chicago

Glander KE (1978) Howling monkey feeding behavior and plant secondary compounds: a study of strategies. In: Montgomery GG (ed) The ecology of arboreal folivores. Smithsonian Institution Press, Washington, DC

Glander KE (1981) Feeding patterns in mantled howling monkeys. In: Kamil A, Sargent T (eds) Foraging behavior: ecological and psychological approaches. Garland, New York

Glanville EJ, Seebacher F (2010) Plasticity in body temperature and metabolic capacity sustains winter activity in a small endotherm (*Rattus fuscipes*). Comp Biochem Physiol A Mol Integr Physiol 155(3):383–91

Glanville EJ, Murray SA, Seebacher F (2011) Thermal adaptation in endotherms: climate and phylogeny interact to determine population-level responses in a wild rat. Funct Ecol. doi:10.1111/j.1365-2435.2011.01933.x

Gluckman PD, Lillycrop KA, Vickers MH, Pleasants AB, Phillips ES, Beedle AS, Burdge GC, Hanson MA (2007) Metabolic plasticity during mammalian development is directionally dependent on early nutritional status. Proc Natl Acad Sci USA 104:12796–12800

Gómez-Gardeñes J, Zamora-López G, Moreno Y, Arenas A (2010) From modular to centralized organization in functional areas of the cat cerebral cortex. PLoS One 5:e12313

Gomulkiewicz R, Holt RD (1995) When does evolution by natural selection prevent extinction? Evolution 49:201–207

Gotthard K, Nylin S (1995) Adaptive plasticity and plasticity as an adaptation: a selective review of plasticity in animal morphology and life history. Oikos 74:3–17

Grant TR, Temple-Smith PD (1998) Field biology of the platypus (*Ornithorhynchus anatinus*): historical and current perspectives. Philos Trans R Soc Lond B 353:1081–1091

Greenfield MD (2002) Signalers and receivers. Oxford University Press, Oxford

Greenstreet SPR, McMillan JA, Armstrong (1998) Seasonal variation in the importance of pelagic fish in the diet of piscivorous fish in the Moray Firth, NE Scotland: a response to variation in prey abundance? ICES J Mar Sci 55:121–133

Gross MR (1996) Alternative reproductive strategies and tactics: diversity within sexes. Trends Ecol Evol 11:92–97

Groves C (1989) A theory of human and primate evolution. Clarendon Press, Oxford

Groves C (2001) Primate taxonomy. Smithsonian Institution Press, Washington, DC

Gustafsson E, Krief S, St. Jalme M (2011) Neophobia and learning mechanisms: how captive orangutans discover medicinal plants. Folia Primatol 82:45–55

Gutman A (1977) Positive contrast, negative induction, and inhibitory stimulus control in rats. J Exp Anal Behav 27:219–233

Haley SL, Lamb JG, Franklin MR, Constance JE, Dearing MD (2007) Xenobiotic metabolism of plant secondary compounds in juniper (*Juniperus monosperma*) by specialist and generalist wood rat herbivores, Genus *Neotoma*. Comp Biochem Physiol C 146:552–560

Hallgrímsson B, Hall BK (2005) Variation: a central concept in biology. Academic, London

Hamilton WD (1964) The evolution of social behavior. J Theor Biol 7:1–52

Hamilton MJ, Davidson AD, Sibley RM, Brown JH (2011) Universal scaling of production rates across mammalian lineages. Proc R Soc Lond 278:560–566

Handford P, Nottebohm J (1976) Allozymic and morphological variation in population samples of rufous-collared sparrow, *Zonotrichia capensis*, in relation to vocal dialects. Evolution 30:802–817

Hanski IA, Gilpin ME (eds) (1997) Metapopulation biology: ecology, genetics, and evolution. Academic, New York

Hanya G, Kiyona M, Hayaishi S (2007) Behavioral thermoregulation of wild Japanese macaques: comparisons between two subpopulations. Am J Primatol 69:802–815

Hau M (2007) Regulation of male traits by testosterone: implications for the evolution of vertebrate life histories. BioEssays 29:133–144

Heidel AJ, Lawal HM, Felder M et al (2011) Phylogeny-wide analysis of social amoeba genomes highlights ancient origins for complex intercellular communication. Genome Res 21:1882–1891

Hirabayashi Y, Gotoh Y (2010) Epigenetic control of neural precursor cell fate during development. Nat Rev Neurosci 11:377–388

Hoelzel AR, Dahlheim M, Stern SJ (1998) Low genetic variation among killer whales (*Orcinus orca*) in the eastern North Pacific and genetic differentiation between foraging specialists. J Heredity 89:121–128

Horwich RH, Brockett RC, James RA, Jones CB (2001) Population structure and group productivity of the Belizean black howling monkey (*Alouatta pigra*): implications for female socioecology. Primate Rep 61:47–65

Houtkooper RH, Argmann C, Houten SM, Canto C, Jeninga EH, Andreux PA, Thomas C, Doenlen R, Schoonjans K, Auwerx J (2011) The metabolic footprint of aging in mice. Sci Rep. doi:10.1038/srep00134

Hsu M, Bhatt M, Adolphs R, Tranel D, Camerer CF (2005) Neural systems responding to degrees of uncertainty in human decision-making. Science 310:1680–1683

Huda A, Jordan IK (2009) Epigenetic regulation of mammalian genomes by transposable elements. Ann NY Acad Sci 1178:276–284

Huey RB, Kingsolver JG (2011) Variation in universal temperature dependence of biological rates. Proc Natl Acad Sci USA 108:10377–10378

Hull CL (1934) Learning II: The factor of the conditioned reflex. In: Murchison C (ed) A handbook of general experimental psychology. Clarke University Press, Worcester, MA

Hunt BG, Ometto L, Wurm Y, Shoemaker D, Yi SV, Keller L, Goodisman MAD (2011) Relaxed selection is a precursor to the evolution of phenotypic plasticity. Proc Natl Acad Sci USA. doi:10.1073/pnas.1104825108

Hunter CM, Caswell H, Runge MC, Regehr EV, Amstrup SC, Stirling I (2010) Climate change threatens polar bear populations: a stochastic demographic model. Ecology 91:2883–2897

Ingram T, Mahler DL (2011) Niche diversification follows key innovation in Antarctic fish radiation. Mol Ecol 20:4590–4591

James RA, Leberg PL, Quattro JM, Vrijenhoek RC (1997) Genetic diversity in black howler monkeys (*Alouatta pigra*) from Belize. Am J Phys Anthropol 102:329–336

Jamieson JG, Mc Rae SB, Trewby M, Simmons RE (2000) High rates of conspecific brood parasitism and egg rejection in coots and moorhens in ephemeral wetlands in Namibia. Auk 117:250–252

Jansen VAA, Stumpf PH (2005) Making sense of evolution in an uncertain world. Science 309:2005–2007

Jerison HJ (1983) The evolution of the mammalian brain as an information-processing system. In: Eisenberg JF, Kleiman DG (eds) Advances in the study of mammalian behavior. American Society of Mammalogists, Shippensburg, PA

Johnstone RA, Cant MA (1999) Reproductive skew and the threat of eviction: a new perspective. Proc R Soc Lond B 266:275–279

Jones CB (1980) The functions of status in the mantled howler monkey (*Alouatta palliata* Gray): intraspecific competition for group membership in a folivorous Neotropical primate. Primates 21:389–405

Jones CB (1982) A field manipulation of social relations among male mantled howler monkeys. Primates 23:130–134

Jones CB (1983) Social organization of captive black [now, black-and-gold: Groves 2001] howler monkeys (*Alouatta caraya*): "social competition" and the use of non-damaging behavior. Primates 24:25–39

Jones CB (1995a) Howler monkeys appear to be preadapted to cope with habitat fragmentation. Endangered Species Update 12:9

Jones CB (1995b) Howler subgroups as homeostatic mechanisms in disturbed habitats. Neotrop Primates 3:7–9

Jones CB (1996a) Predictability of plant food resources for mantled howler monkeys at Hacienda La Pacifica, Costa Rica: Glander's dissertation revisited. Neotropical Primates 4:147–149

Jones CB (1996b) Relative reproductive success in the mantled howler monkey: implications for conservation. Neotrop Primates 4:21–23

Jones CB (1997a) Life history patterns of howler monkeys in a time-varying environment. Boletin Primatologico Latinoamericano 6:1–8

Jones CB (1997b) Rarity in primates: implications for conservation. Mastozoología Neotropical 4:35–47

Jones CB (1997c) Social parasitism in the mantled howler monkey, *Alouatta palliata* Gray (Primates: Cebidae [now Atelidae]: involuntary altruism in a mammal? Sociobiology 30:51–61

Jones CB (1997d) Subspecific differences in vulva sizes between *Alouatta palliata palliata* and *A. p. mexicana*: implications for assessment of female receptivity. Neotrop Primates 5:46–48

Jones CB (1999) Testis symmetry in the mantled howler monkey. Neotrop Primates 7:117–119

Jones CB (2002) Genital displays by adult male and female mantled howling monkeys, *Alouatta palliata* (Atelidae): evidence for condition-dependent compound displays. Neotrop Primates 10:144–147

Jones CB (2005a) Behavioral flexibility in primates: causes and consequences. Springer, New York

Jones CB (2005b) Phenotype as developmental bridge: whither nature and nurture? Am J Psychol 118:141–158, book review of West-Eberhard 2003

Jones CB (2005c) Social parasitism in mammals with particular reference to neotropical primates. Mastozoología Neotropical 12:19–35

Jones CB (2006) An exploratory analysis of developmental plasticity in Costa Rican mantled howler monkeys (*Alouatta palliata palliata*). In: Estrada A, Garber PA, Pavelka M, Luecke L (eds) New perspectives in the study of Mesoamerican primates: distribution, ecology, behavior, and conservation. Springer, New York

Jones CB (2008) Ethology, neuroethology, and evolvability in vertebrates: a brief review and prospectus. Primate Rep 75:41–62

Jones CB (2009) The effects of heterogeneous regimes on reproductive skew in eutherian mammals. In: Hager R, Jones CB (eds) Reproductive skew in vertebrates: proximate and ultimate causes. Cambridge University Press, Cambridge

Jones CB, Agoramoorthy G (2003) Alternative reproductive behaviors in primates: towards general principles. In: Jones CB (ed) Sexual selection and reproductive competition in primates: new perspectives and directions. American Society of Primatologists, Norman

Jones CB, van Cantfort TE (2007) Multimodal communication by male mantled howler monkeys (*Alouatta palliata*) in sexual contexts: a descriptive analysis. Folia Primatol 78:166–185

Jordan IK (2006) Evolutionary tinkering with transposable elements. Proc Natl Acad Sci USA 103:7941–7942

Kandel E, Schwartz J, Jessell T (2000) Principles of neuroscience. McGraw-Hill, New York

Kaneko K (2009) Relationship among phenotypic plasticity, phenotypic fluctuations, robustness, and evolvability; Waddington's legacy revisited under the spirit of Einstein. J Biosci 34:529–542

Katsnelson E, Motro U, Feldman MW, Lotem A (2011) Evolution of learned strategy choice in a frequency-dependent game. Proc R Soc Lond B. doi:10.1098/rslb.2011.1734

Kawai M (1965) Newly acquired pre-cultural behavior of the natural troop of Japanese monkeys on Koshima Islet. Primates 1:1–30

Keane TM, Goodstadt L, Danecek P, White MA, Wong K, Yalcin B, Heger A, Agam A et al (2011) Mouse genomic variation and its effect on phenotypes and gene regulation. Nature 477:289–294

Keller L (1995) Social life: the paradox of multiple-queen colonies. Trends Ecol Evol 10:355–360

Kelm DH, Simon R, Kuhlow D, Voigt CC, Ristow M (2011) High activity enables life on a high-sugar diet: blood glucose regulation in nectar-feeding bats. Proc R Soc Lond B. doi:10.1098/rspb.2011.0465

Kermack DM, Kermack KA (1984) Evolution of mammalian characters. Croom Helm, London

King HM, Shubin NH, Coates MI, Hale ME (2011) Behavioral evidence for the evolution of walk-ing and bounding before terrestriality in sarcopterygian fishes. Proc Natl Acad Sci USA. http://www.pnas.org/cgi/doi/10.1073/pnas.1118669109

Kirschner M, Gerhart J (1998) Evolvability. Proc Natl Acad Sci 95:8420–8427

Kishida KT, Sandberg SG, Lohrenz T, Comair YG, Sáez I, Phillips PEM, Montague PR (2011) Sub-second dopamine detection in human striatum. PLoS One 6:e23291. doi:10.1371/journalpone.0023291

Kitazoe Y, Kishino H, Hasegawa M, Matsui A, Land N, Tanaka M (2011) Stability of mitochon-drial membrane proteins in terrestrial vertebrates predicts aerobic capacity and longevity. Genome Biol Evol 3:1233–1244

Korb E, Finkbeiner S (2011) Arc in synaptic plasticity: from gene to behavior. Trends Neurosci 34:591–598

Krause EG, de Kloet AD, Flak JN, Smeltzer MD, Solomon MB, Evanson NK, Woods SC, Sakai RR, Herman JP (2011) Hydration state controls stress responsiveness and social behavior. Neuroscience 31:5470–5476

Krebs CJ, Cowcill K, Boonstra R, Kenney AJ (2010) Do changes in berry crops drive population fluctuations in small rodents in the southwestern Yukon? J Mammal 91:500–509

Krützen M, Mann J, Heithaus MR, Connor RC, Bejder L, Sherwin WB (2005) Cultural transmis-sion of tool use in bottlenose dolphins. Proc Natl Acad Sci USA 102:8939–8943

Kuraku S, Kuratani S (2011) Genome-wide detection of gene extinction in early mammalian evo-lution. Genome Biol Evol 3:1449–1462. doi:10.1093/gbe/evr120

Kussell E, Leibler S (2005) Phenotypic diversity, population growth, and information in fluctuating environments. Science 309:2075–2078

Lande R (2009) Adaptation to an extraordinary environment by evolution of phenotypic plasticity and genetic assimilation. J Evol Biol 22:1435–1446

Laughlin SB, Sejnowski J (2003) Communication in neural networks. Science 301:1870

Lee HK, Arbarosie M, Kameyama K, Bear MF, Huganir RL (2000) Regulation of distinct AMPA receptor phosphorylation sites during bidirectional synaptic plasticity. Nature 405:955–959

Lee CE, Remfert JL, Chang YM (2006) Response to selection and evolvability of invasive popula-tions. Genetica 129:179–192

Lerner M (1954) Genetic homeostasis. Dover, New York

Lewontin RC (1957) The adaptations of populations to varying environments. Cold Spring Harb Symp Quant Biol 22:395–408

Libert C, Dejager L, Pinheiro I (2010) The X-chromosome in immune functions: when a chromo-some makes the difference. Nat Rev Immunol 10:594–604

Loewenstein Y, Seung HS (2006) Operant matching is a generic outcome of synaptic plasticity based on the covariance between reward and neural activity. Proc Natl Acad Sci USA 103:15224–15229

Lorenz K (1981) The foundations of ethology. Springer, Berlin

Maan ME, Seehausen O (2011) Ecology, sexual selection and speciation. Ecol Lett. doi:10.1111/j.1461-0248.2011.01606.x

MacDonald DW, Johnson DDP (2001) Dispersal in theory and practice: consequences for popula-tion biology. In: Clobert J, Danchin E, Dhondt AA, Nichols JD (eds) Dispersal. Oxford University Press, Oxford

Malmgren LA (1979) Empirical population genetics of golden mantled howling monkeys (*Alouatta palliata*) in relation to population structure, social dynamics, and evolution. Unpublished Ph.D. Dissertation, University of Connecticut, Storrs

Maren S (1999) Long-term potentiation in the amygdala: a mechanism for emotional learning and memory. TINS 22:561–567

Matzel LD, Han YR, Grossman H (2003) Individual differences in the expression of a "general" learning ability in mice. J Neurosci 23:6423–6433

Maynard Smith J, Harper D (2003) Animal signals. Oxford University Press, Oxford

Mayr E (1976) Evolution and the diversity of life. Harvard University Press, Cambridge

Mayr E (1982) The growth of biological thought: diversity, evolution, and inheritance. Harvard University Press, Cambridge

Mazur JE (2004) Risky choice: selecting between certain and uncertain outcomes. Behav Anal Today 5:190–203

McCleery RG (1978) Optimal behavior sequences and decision-making. In: Krebs JR, Davies NB (eds) Behavioural ecology: an evolutionary approach. Sinauer, Sunderland, MA

McCoy MW, Bolker BM, Warkentin KM, Vonesh JR (2011) Predicting predation through prey ontogeny using size-dependent functional response models. Am Nat 177:752–766

McNab B (1974) The energetics of endotherms. Ohio J Sci 74:370–380

McNab B (1980) Food habits, energetics, and the population biology of mammals. Am Nat 116:106–124

McNab B (1986) The influence of food habits on the energetics of eutherian mammals. Ecol Monogr 56:1–19

McNab B (2005) Uniformity in the basal metabolic rate of marsupials: its causes and consequences. Revista Chilena de Historia Natural 78:183–198

McNab B (2006) The energetics of reproduction in endotherms and its implication for their conservation. Integr Comp Biol 46:1159–1168

Meachen-Samuels JA (2012) Morphological convergence of the prey-killing arsenal of sabertooth predators. Paleobiology 38:1–14

Meachen-Samuels J, Van Valkenburgh B (2009) Forelimb indicators of prey-size preference in the Felidae. J Morphol 270:729–744

Meiklejohn CD, Hartl DL (2002) A single mode of canalization. Trends Ecol Evol 17:468–473

Meredith RW, Janečka JE, Gatesy J, Ryder OA, Fisher CA, Teeling EC, Goodbla A, Eizirik E et al (2011) Impacts of the Cretaceous terrestrial revolution and KPg extinction on mammal diversification. Science 334:521–524

Meyer M, Plass M, Pérez-Valle J, Eyras E, Vilardell J (2011) Deciphering 3'ss selection in the yeast genome reveals an RNA thermosensor that mediates alternative splicing. Mol Cell 43:1033–1039

Miller W, Hayes VM, Ratan A, Petersen DC, Wittekindt NE, Miller J, Walenz B, Knight J et al (2011) Genetic diversity and population structure of the endangered marsupial *Sarcophilus harrisii* (Tasmanian devil). Proc Natl Acad Sci USA 108:12348–12353

Møller AP, Pomiankowski A (1994) Fluctuating asymmetry and sexual selection. In: Markow T (ed) Developmental instability: its origins and evolutionary implications. Kluwer Academic, Dordrecht

Moore J (2002) Parasites and the behavior of animals. Oxford University Press, Oxford

Morris D (1957) "Typical intensity" and its relation to the problem of ritualization. Behaviour 11:1–12

Nabholz B, Mauffrey J-F, Bazin E, Galtier N, Glémin S (2008) Determination of mitochondrial genetic diversity in mammals. Genetics 178:351–361

Nevo E (2001) Evolution of genome-phenome diversity under environmental stress. Proc Natl Acad Sci USA 98:6233–6240

Nijhout HF (2003a) Development and evolution of adaptive polyphenisms. Evol Dev 5:9–18

Nijhout HF (2003b) The control of growth. Development 130:5863–5867

Nilsonne G, Applegren A, Axelson J, Fredrikson LM (2011) Learning in a simple biological system: a pilot study of classical conditioning of human macrophages *in vitro*. Behav Brain Funct 7:47. doi:10.1186/1744-9081-7-47

Nosil P, Crespi BJ (2006) Experimental evidence that predation promotes divergence in adaptive radiation. Proc Natl Acad Sci USA 103:9090–9095

Nugent BM, McCarthy MM (2011) Epigenetic underpinnings of developmental sex differences in the brain. Neuroendocrinology 93:150–158

O'Brien SJ, Menotti-Raymond M, Murphy WJ, Nash WG, Weinberg J, Stanyon R, Copeland NG, Jenkins NA, Womack JE, Marshall Graves JA (1999) The promise of comparative genomics in mammals. Science 286:458–481

Padykula HA, Taylor JM (1982) Marsupial placentation and its evolutionary significance. J Reprod Fertil 31:95–104

Pan Y, Tsai C-J, Ma B, Nussinov R (2010) Production of ATP regulated by feedback effects. Trends Genet 26:75–83

Park T, Lloyd M (1955) Natural selection and the outcome of competition. Am Nat 89:235–240

Parker GA (1974) Assessment strategy and the evolution of fighting behaviour. J Theor Biol 47:223–243

Pasque V, Jullien J, Miyamoto K, Halley-Stott RP, Gurdon JB (2011) Epigenetic factors influencing resistance to nuclear reprogramming. Trends Genet 27:516–525

Pavlov IP (1927) Conditioned reflexes. Oxford University Press, Oxford

Pavlov IP (1928) Lectures on conditioned reflexes. International, New York

Perez-Tomé JM, Toro MA (1982) Competition of similar and non-similar genotypes. Science 299:153–154

Piersma T, Drent J (2003) Phenotypic flexibility and the evolution of organismal design. Trends Ecol Evol 18:228–233

Pigliucci M (1996) How organisms respond to environmental changes: from phenotypes to molecules (and vice versa). Trends Ecol Evol 11:168–173

Pigliucci M (2008) Is evolvability evolvable? Nat Rev Genet 9:75–82

Pigliucci M, Muller GB (eds) (2010) Evolution: the extended synthesis. MIT Press, Cambridge

Popoli M, Yan Z, McEwen BS, Sanacora G (2011) The stressed synapse. Nat Rev Neurosci 13:22–37

Porter WP, Kearney M (2009) Size, shape, and the thermal niche of endotherms. Proc Natl Acad Sci USA 106:19666–19672

Poulin R (2002) Parasite manipulation of host behavior. In: Lewis EE, Campbell JF, Sukhedeo MVK (eds) The behavioral ecology of parasites. CABI Publishing, New York, NY

Price TD (2006) Phenotypic plasticity, sexual selection, and the evolution of colour patterns. J Exp Biol 209:2368–2376

Proppe DS, Sturdy CB, St. Clair CC (2011) Flexibility in animal signals facilitates adaptation to rapidly changing environments. PLoS One 6:e25413

Pulliam HR, Caraco T (1984) Living in groups: Is there an optimal group size? In: Krebs JR, Davies NB (eds) Behavioural ecology: an evolutionary approach. Sinauer Associates, Sunderland, MA

Pyenson ND, Lindberg DR (2011) What happened to gray whales during the Pleistocene? The ecological impact of sea-level change on benthic feeding areas in the north Pacific Ocean. PLoS One 6:e21295

Quattara K, Lemasson A, Zuberbuhler K (2009) Campbell's monkeys concatenate vocalizations into context-specific call sequences. Proc Natl Acad Sci USA 106:22026–22031

Rajakumar R, San Mauro D, Dijkstra MB, Huang MH, Wheeler DE, Hiou-Tim F, Khila MH, Cournoyea M, Abouheif E (2012) Ancestral developmental potential facilitates parallel evolution in ants. Science 335:79–82

Rannala BH, Brown CR (1994) Relatedness and conflict over optimal group size. Trends Ecol Evol 9:117–118

Réale D, McAdam AG, Boutin S, Berteaux D (2003) Genetic and plastic responses of a northern mammal to climate change. Proc R Soc Lond 270:591–596

Reed TE, Schindler DE, Waples RS (2011) Interacting effects of phenotypic plasticity and evolution on population persistence in a changing climate. Conserv Biol 25:56–63

Reznick DA, Bryga H, Endler JA (1990) Experimentally induced life-history evolution in a natural population. Nature 346:357–359

Rice WR (2000) Dangerous liaisons. Proc Natl Acad Sci USA 97:12953–12955

Ricklefs RE (1977) On the evolution of reproductive strategies in birds: reproductive effort. Am Nat 111:453–478

Ricklefs RE, Wikelsky M (2002) The physiology/life-history nexus. Trends Ecol Evol 17:462–468

Rodriguez-Cabal MA, Branch LC (2011) Influence of habitat factors on the distribution and abundance of a marsupial seed disperser. J Mammal 92:1245–1252

Roff D (2002) Life history evolution. Sinauer, Sunderland, MA

Rosenzweig MR (2007) Modification of brain circuits through experience. In: Bermúdez-Rattoni F (ed) Neural plasticity and memory: from genes to brain imaging. CRC, Boca Raton, FL

Rowe L, Arnqvist G (2012) Sexual selection and the evolution of genital shape and complexity in water striders. Evolution 66:40–54

Roylance D (2000) Introduction to elasticity. http://freeit.free.fr/elasticity/david%20roylance%20-%20mechanics%20of%20materials.pdf

Roylance D (2001) Stress-strain curves. http://ocw.mit.edu/courses/materials-science-and-engineering/3-11-mechanics-of-materials-fall-1999/modules/ss.pdf

Rutherford SL (2000) From genotype to phenotype: buffering mechanisms and the storage of genetic information. BioEssays 22:1095–1105

Saltzman W (2003) Reproductive competition among female common marmosets (*Callithrix jacchus*): proximate and ultimate causes. In: Jones CB (ed) Sexual selection and reproductive competition in primates: new perspectives and directions. American Society of Primatologists, Norman, OK

Sargeant BL, Wirsing AJ, Heithaus MR, Mann J (2007) Can environmental heterogeneity explain individual foraging variation in wild bottlenose dolphins (*Tursiops* sp.)? Behav Ecol Sociobiol 61:679–688

Savolainen R, Vespalainen K (2003) Sympatric speciation through intraspecific social parasitism. Proc Natl Acad Sci USA 100:7169–7174

Schoener TW (1971) Theory of feeding strategies. Annu Rev Ecol Syst 2:369–404

Schoener TW (2011) The newest synthesis: understanding the interplay of evolutionary and ecological dynamics. Science 331:426–429

Schultz W (2006) Behavioral theories and the neurophysiology of reward. Annu Rev Psychol 57:87–115

Schwartz OA, Armitage KB (1980) Genetic variation in social mammals: the marmot model. Science 207:665–667

Scott-Phillips TC, Blythe RA, Gardner A, West SA (2012) How do communication systems emerge? Proc R Soc Lond B. doi:10.1098/rspb.2011.2181

Seebacher F, Glanville EJ (2010) Low levels of physical activity increase metabolic responsiveness to cold in a rat (*Rattus fuscipes*). PLoS One 5(9):e13022

Seebacher F, Brand MD, Else PL, Guderley H, Hulbert AJ, Moyes CD (2010) Plasticity of oxidative metabolism in variable climates: molecular mechanisms. Physiol Biochem Zool 83:721–732

Selander RK, Kaufman DW (1973) Genic variability and strategies of adaptation in animals. Proc Natl Acad Sci USA 70:1875–1877

Servedio M (2001) Beyond reinforcement: the evolution of premating isolation by direct selection on preferences and postmating prezygotic incompatibilities. Evolution 55:1909–1920

Seyfarth RM, Cheney DL, Marler P (1980) Monkey responses to three different alarm calls: evidence of predator classification and semantic communication. Science 210:801–803

Shafer AB, Fan CW, Côté SD, Coltman DW (2012) Lack of genetic diversity in immune genes predates glacial isolation in the North American mountain goat (*Oreamnos americanus*). J Heredity. doi:1093/jhered/esr138

Sharma RC, Inoue S, Roitelman J, Schimke RT, Simoni RD (1992) Peptide transport by the multidrug resistance pump. J Biol Chem 267:5731–5734

Shelford VE (1911) Ecological succession. Marine Biological Laboratory, Woods Hole, MA

Siegal ML, Bergman A (2002) Waddington's canalization revisited: developmental stability and evolution. Proc Natl Acad Sci USA 99:10528–10532

Sih A (1985) Predation, competition, and prey communities: a review of field experiments. Annu Rev Ecol Systemat 16:292–311

Skinner BF (1935) The generic nature of the concepts of stimulus and response. J Gen Psychol 12:40–65

Skinner BF (1981) Selection by consequences. Science 213:501–504

Slobodkin LB, Rapoport A (1974) An optimal strategy of evolution. Q Rev Biol 49:181–200

Sniegowski PD, Murphy HA (2006) Evolvability. Curr Biol 16:R831–R834

Somel M, Liu X, Tang L, Yan Z, Hu H, Guo S, Jiang X, Zhang X, Xu G, Xie G, Li N, Hu Y, Chen W, Pääbo S, Khaitovich P (2011) MicroRNA-driven developmental remodeling in the brain distinguishes humans from other primates. PLoS Biol 9:e1001214

Soria-Carrasco V, Castresana J (2011) Patterns of mammalian diversification in recent evolutionary times: global tendencies and methodological issues. J Evol Biol 24:2611–2623

Sporns O (2011) Networks of the brain. MIT Press, Cambridge, MA

St. John Smith E, Omerbašić D, Lechner SG, Anirudhan G, Łapatsina L, Lewin GR (2011) The molecular basis of acid insensitivity in the African naked mole-rat. Science 334:1557–1560

Stearns SC (1992) The evolution of life histories. Oxford University Press, Oxford

Stearns SC (2000) Life history evolution: successes, limitations, and prospects. Naturwissenschaften 87:476–486

Stearns SC, Koella JC (1986) The evolution of phenotypic plasticity in life-history traits: predictions of reaction norms for age and size at maturity. Evolution 40:893–913

Stearns SC, de Jong G, Newman B (1991) The effects of phenotypic plasticity on genetic correlations. Trends Ecol Evol 6:122–126

Sterck F, Markesteijn L, Schieving F, Poorter L (2011) Functional traits determine trade-offs and niches in a tropical forest community. Proc Natl Acad Sci USA. http://www.pnas.org/cgi/doi/10.1073/pnas.1106950108

Stevenson PR, Castellanos MC, Cortés AI, Link A (2008) Flowering patterns in a seasonal tropical lowland forest in Western Amazon. Biotropica 40:559–567

Strassmann JE, Queller DC (2011) Evolution of cooperation and control of cheating in a social microbe. Proc Natl Acad Sci USA 108:10855–10862

Strobeck C (1975) Selection in a fine-grained environment. Am Nat 109:414–426

Suggitt AJ, Gillingham PK, Hill JK, Huntley B, Kunin WE, Roy DB, Thomas CD (2011) Habitat microclimates drive fine-scale variation in extreme temperatures. Oikos 120:1–8

Sultan SE, Spencer HG (2002) Metapopulation structure favors plasticity over local adaptation. Am Nat 160:271–283

Swanson WJ, Yang Z, Wolfner MF, Aquadro CF (2001) Positive Darwinian selection drives the evolution of several female reproductive proteins in mammals. Proc Natl Acad Sci USA 98:2509–2514

Swedell L, Schrier A (2009) Male aggression toward females in Hamadryas baboons: conditioning, coercion, and control. In: Mubler MN, Wrangham RW (eds) Sexual coercion in primates and humans: an evolutionary perspective on male aggression against females. Harvard University Press, Cambridge, MA

Templeton AR, Rothman ED (1974) Evolution in heterogeneous environments. Am Nat 108:409–428

Terblanche JS, Hoffmann MKA, Rako L, le Roux PC, Chown SL (2011) Ecologically relevant measures of tolerance to potentially lethal temperatures. J Exp Biol 214:3713–3725

Thorndike EL (1898) Animal intelligence: an experimental study of the associative processes in animals. Psychol Rev 73:16–43

Thorndike EL (1911) Animal intelligence. Macmillan, New York

Tills O, Rundle SD, Salinger M, Ha MNT, Spicer JI (2011) A genetic basis for intraspecific differences in developmental timing? Evol Dev 13:542–548

Tinbergen N (1952) Derived activities: their causation, biological significance, origin, and emancipation during evolution. Q Rev Biol 27:1–32

Trivers RL (1972) Parental investment and sexual selection. In: Campbell B (ed) Sexual selection and the descent of man 1871-1971. Aldine, New York

Tsankov A, Yanagisawa Y, Rhind N, Regev A, Rando OJ (2011) Evolutionary divergence of intrinsic and *trans*-regulated nucleosome positioning sequences reveals plastic rules for chromatin organization. Genome Res 21:1851–1862

Urban MC, Tewksbury JJ, Sheldon KS (2011) On a collision course: competition and dispersal differences create no-analogue communities and cause extinctions during climate change. Proc R Soc Lond B. doi:10.1098/rspb.2011.2367

Van Milgen J (2002) Modeling biochemical aspects of energy metabolism in mammals. J Nutr 132:3195–3202

Vaughan TA (1972) Mammalogy. W.B. Saunders, Philadelphia, PA

Vaughan TA, Ryan JM, Czaplewski NJ (2000) Mammalogy, 4th Edition. Brooks/Cole, USA

Venditti C, Meade A, Pagel M (2011) Multiple routes to mammalian diversity. Nature. doi:10.1038/nature10516

Vickery TJ, Chun MM, Lee D (2011) Ubiquity and specificity of reinforcement signals throughout the human brain. Neuron 72:166–177

Vieira-Silva S, Touchon M, Abby SS, Rocha EPC (2011) Investment in rapid growth shapes the evolutionary rates of essential proteins. Proc Natl Acad Sci USA. http://www.pnas.org/cgi/doi/10.1073/pnas.1110972108

Villalobos F, Valerio AA, Retana AP (2004) A phylogeny of Howler monkeys (Cebidae: *Alouatta*) based on mitochondrial, chromosomal and morphological data. Rev Biol Trop 52:665–677

Vogels R, de Graaff W, Deschamps J (1990) Expression of the murine homeobox-containing gene Hox-2.3 suggests multiple time-dependent and tissue-specific roles during development. Development 110:1159–1168

Wacongne C, Labyt E, van Wassenhove V, Bekinschtein T, Naccache L, Dehaene S (2011) Evidence for a hierarchy of predictions and prediction errors in human cortex. Proc Natl Acad Sci USA. http://www.pnas.org/cgi/doi/10.1073/pnas.1117807108

Wagner GP (ed) (2001) The character concept in evolutionary biology. Academic, San Diego

Wagner A (2008) Robustness and evolvability: a paradox resolved. Proc R Soc Lond B 275:91–100

Wagner A (2012) The role of robustness in phenotypic adaptation and innovation. Proc R Soc Lond B. doi:10.1098/rspb.2011.2293

Wang Z, Gerstein M, Snyder M (2009) RNA-sequencing: a revolutionary tool for transcriptomics. Nat Rev Genet 10:57–63

Wasserman EA (1973) Pavlovian conditioning with heat reinforcement produces stimulus-directed pecking in chicks. Science 81:875–877

Weinreich DM, Delaney NF, DePristo MA, Hartl DL (2006) Darwinian evolution can follow only very few mutational paths to fitter proteins. Science 312:111–114

Weisbecker V, Goswami A (2010) Brain size, life history, and metabolism at the marsupial/placental dichotomy. Proc Natl Acad Sci 107:16216–16221

Weiss LA, Pan L, Abney M, Ober C (2006) The sex-specific genetic architecture of quantitative traits in humans. Nat Genet 38:218–222

West SA, Pen I, Griffin AS (2002) Cooperation and competition between relatives. Science 296:72–75

West-Eberhard MJ (2003) Developmental plasticity and evolution. Oxford University Press, New York

West-Eberhard MJ (2005) Phenotypic accommodation: adaptive innovation due to developmental plasticity. J Exp Zool 304B:610–618

Wheeler WM (1928) Insect societies: their origins and evolution. Harcourt Brace, New York

White RJ, Sharrocks AD (2010) Coordinated control of the gene expression machinery. Trends Genet 26:214–220

Wilkie CO (2011) Transcriptional robustness complements nonsense-mediated decay in humans. PLoS Genet 7:31002296

Wilson EO (1971) The insect societies. Cambridge, Belknap

Wilson EO (1975) Sociobiology. Cambridge, Belknap

Wilson DE, Reeder DM (eds) (2005) Mammal species of the world, 2 vols. Johns Hopkins University Press, Baltimore, MD

Wittenberger JF (1980) Group size and polygamy in social mammals. Am Nat 115:197–222

Woodard SH, Fischman BJ, Venkat A, Hudson ME, Varala K, Cameron SA, Clark AG, Robinson GE (2011) Genes involved in convergent evolution of eusociality in bees. Proc Natl Acad Sci USA. http://www.pnas.org/cgi/doi.1073/pnas.1103457108

Woods SC, Kulkosky PJ (1976) Classically conditioned changes of blood glucose level. Psychosom Med 38:201–219

Woolley L-A, Page B, Slotow R (2011) Foraging strategy within African elephant family units: why body size matters. Biotropica 43:489–495

Wright S (1949) Adaptation and selection. In: Jepsin GC, Mayr E, Simpson GG (eds) Genetics, paleontology, and evolution. Princeton University Press, Princeton

Wu D-D, Zhang Y-P (2011) Different level of population differentiation among human genes. BMC Evol Biol. http://www.bomedcentral.com/1471-2148/11/16

Zhu L, Wu Q, Dai J, Zhang S, Wei F (2011) Evidence of cellulose metabolism by the giant panda gut microbiome. Proc Natl Acad Sci USA. doi:10.1073/pnas/017956108

Zink AG, Reeve HK (2005) Predicting the temporal dynamics of reproductive skew and group membership in communal breeders. Behav Ecol 16:880–888. doi:10,1093/beheco/ari062

Index

C.B. Jones, *Robustness, Plasticity, and Evolvability in Mammals: A Thermal Niche
Approach*, SpringerBriefs in Evolutionary Biology, DOI 10.1007/978-1-4614-3885-4,
© Clara B. Jones 2012